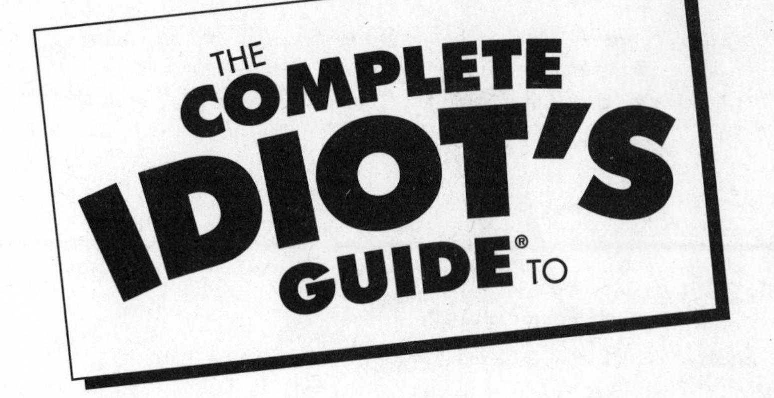

Biology

by Glen E. Moulton, Ed.D.

A member of Penguin Group (USA) Inc.

For my wife, Kimberly, who is the springtime of my life, and my children, Grant, Lauren, and Jessica, who constitute the rest of the seasons. They provide me with overflowing joy. They also helped write this book.

ALPHA BOOKS

Published by the Penguin Group

Penguin Group (USA) Inc., 375 Hudson Street, New York, New York 10014, U.S.A.

Penguin Group (Canada), 10 Alcorn Avenue, Toronto, Ontario, Canada M4V 3B2 (a division of Pearson Penguin Canada Inc.)

Penguin Books Ltd, 80 Strand, London WC2R 0RL, England

Penguin Ireland, 25 St Stephen's Green, Dublin 2, Ireland (a division of Penguin Books Ltd)

Penguin Group (Australia), 250 Camberwell Road, Camberwell, Victoria 3124, Australia (a division of Pearson Australia Group Pty Ltd)

Penguin Books India Pvt Ltd, 11 Community Centre, Panchsheel Park, New Delhi—110 017, India

Penguin Group (NZ), Cnr Airborne and Rosedale Roads, Albany, Auckland, New Zealand (a division of Pearson New Zealand Ltd)

Penguin Books (South Africa) (Pty) Ltd, 24 Sturdee Avenue, Rosebank, Johannesburg 2196, South Africa

Penguin Books Ltd, Registered Offices: 80 Strand, London WC2R 0RL, England

International Standard Book Number: 1-59257-258-8
Library of Congress Catalog Card Number: 2004110007

06 05 04 8 7 6 5 4 3 2 1

Interpretation of the printing code: The rightmost number of the first series of numbers is the year of the book's printing; the rightmost number of the second series of numbers is the number of the book's printing. For example, a printing code of 04-1 shows that the first printing occurred in 2004.

Printed in the United States of America

Publisher: *Marie Butler-Knight*
Product Manager: *Phil Kitchel*
Senior Managing Editor: *Jennifer Chisholm*
Senior Acquisitions Editor: *Mike Sanders*
Development Editor: *Nancy D. Lewis*
Senior Production Editor: *Billy Fields*
Copy Editor: *Keith Cline*
Illustrator: *Richard King*
Cover/Book Designer: *Trina Wurst*
Indexer: *Julie Bess*
Layout: *Angela Calvert*
Proofreader: *John Etchison*

Contents at a Glance

Contents

Foreword

I believe that everyone is a biologist. If you've ever gone on a diet, that's biology! If you've watched a bat swooping for insects at dusk, that's biology. If you've been amused by bendy celery that's been in the refrigerator too long, that's biology too. Amazing plants and animals surround us, and it's only natural to have questions about how they live. While people have an innate ability to distinguish living things from nonliving things, few people can define what it is to be "alive."

The fact that you're reading this suggests that you already have an interest in biology and are seeking a good introduction to the subject. I think *The Complete Idiot's Guide to Biology* is a very good choice for you. However, if you're on the fence, let me make my case for why this book is your best bet.

Biology is booming and discoveries are coming at breakneck speed. I'm reminded of the scene from the classic 1967 movie "The Graduate," in which a savvy family friend sidles up to the young Dustin Hoffman and gives him the one-word, surefire advice he needs for success in the modern business world: "Plastics." If that movie had been filmed in the early eighties the word probably would have been "Computers." And today that word would certainly be "Biology."

But this example gives the impression that the main reason to learn biology is to follow developments in the biotechnology business. The reasons why a good understanding of biology is important go far beyond that. These days, many hot topics are closely related to biology. Diet, disease, behavior, genetic engineering, the environment … trying to understand any of these without knowledge of biology is like trying to learn how a car works without looking under the hood.

Breakthroughs in biology, specifically the biology of DNA, have given us the ability to manipulate crops, livestock, and even our own health in powerful ways. Awareness of the human impact on the environment is also at an all-time high. People are beginning to ask tough ethical questions about the relationship between humans and nature. Regardless of whether you're a student, parent, teacher, businessperson, or politician, the more biology you know the better off you'll be. I applaud every effort to make biology entertaining and accessible.

And of course, don't overlook the basic fact that we humans are *animals* with cells, organs, and behavior. To truly understand ourselves, we must understand our biology. The human body doesn't come with an owner's manual. Luckily, we have *The Complete Idiot's Guide to Biology!*

Biology (studying it, I mean) came pretty naturally to me. But not everyone has the same experience. Biology students often run into trouble at some point during their education, saying, "I hate Genetics," or "Biochemistry is really hard," or "Do I really need to take Ecology?!" In school, biology gets broken up into different categories, making it difficult for students to see the overall picture. What's helpful about this book is that it never loses sight of that bigger picture even as each facet of biology is examined. You see the forest *and* the trees.

You will see that evolution pops up in almost every chapter of this book, and with good reason. Evolution is a central concept in biology, and most biologists agree that whether they are studying cells, chromosomes, organisms, or ecosystems, evolution is at the heart of the questions they seek to answer. Look to the "Bionotes" spread through the chapters for glimpses of the variety that makes biology so entertaining.

Studying biology gives us some important insights. We learn that the differences between humans groups, some of which are used to justify war, are essentially non-existent below the skin. We learn that the genetic material, DNA, is a gift passed from generation to generation back in time from the first life on the planet. And we learn that cells (yours, the cells of the apple you're eating, and the microbes in your gut that will help you digest it) are more alike than different. So whether you're a single-celled bacteria living in a Yosemite hot spring or a multi-cellular eukaryote of the Hominidae family (which is much more likely) you will recognize yourself in this book!

Edward Himelblau
Professor of genetics and cell biology
Long Island University

Introduction

I have helped myself and others learn biology for a very long time. Through several interesting vagabond teaching experiences, I learned to work effectively with a wide range of students. Curiously, I was most attracted to the ones who were most like me—they struggled to learn biology.

In working with students who did not readily comprehend or sometimes even want to readily comprehend biology, I learned several things. Foremost, they all have an unpleasant science experience in their background. It could be something as simple as a weak teacher who was overpowering or simply got confused in the content. I remember one student who arrived in my college zoology class with so much bad information that it took most of the semester to untrain the student and provide her with the correct content to keep her going. However, after a few of the major concepts were clarified, the world of living things became more real and understandable to her. By the time she finished the following botany course, she was equal with her peers.

In other cases, it may have also been an improper sequencing of the material that was inappropriate for the developmental stage of the child. I once had a student who hated science because she had such a tough time with introductory physics. As it turned out, the student did not have the prerequisite mathematics background to support the physics instruction. No wonder she was lost. In some cases, the fear of science has prevented students from succeeding. I remember a student who entered science class for the first time scared sick because he had heard from his older sister that biology class was rough, tough, with limited chance of survival. At first the student struggled, until he got a few successful experiences behind him, and then he was able to move forward.

I have also observed that there are sometimes fallacies in students' understanding of how to study science. After the fear factor and mental blocks are removed, there still remains the question of how to best approach the subject of biology. How does one maximize understanding with the minimal amount of effort and time? Interesting question, and there are several things that every student should do when learning science.

Science is not that hard. It may be difficult because of inappropriate study habits. Try this: Decide what it is that you are supposed to learn. Although this sounds silly, the number-one problem with student understanding is that they do not have a clear understanding of what they are to learn or to what degree they are to learn it, or how they will be assessed to see whether they have learned it. For instance, does your instructor publish a list of objectives, or study guides, or sample test questions? If so, use them to focus your studying efforts. Most enlightened teachers tell students exactly what they are supposed to learn and then help them to learn it through classroom and laboratory sessions. Most even have office hours during which they handle one-on-one situations. Use them, you are paying for it!

Science is a vocabulary unto itself. Simply reading the words is generally not enough. Sometimes the words must be written first and visualized later. Probably the best way to learn vocabulary is not to memorize the words, but to think about their action. What is the word describing? How does it fit into the bigger picture?

Science is an assimilation of simple scientific truths. The most complex processes and ideas are built on a few principles. It has been said that all mathematics can be reduced to addition, subtraction, multiplication, and division. In the same way, biology can be simplified into a few processes that are added together to make bigger, more complicated situations. When confronted with challenging problems, try breaking them down into smaller units. Usually they fall apart into a series of sequential steps that make sense and are easier to learn and remember. For example, the process of photosynthesis can be reduced to the simple concepts of osmosis, diffusion, and a few simple chemical reactions. Divide and conquer!

Please remember that reading science is not like reading a novel. You may have to slow down, read it several times, read a line or phrase at a time, read forward, stop, and go back. Don't feel bad or be embarrassed if you are like me and have had to read it numerous times in many different ways to understand it fully.

This book is designed to help students who need a little assistance to clarify issues and to provide a more complete understanding of scientific and biological principles. Complex material has been simplified without losing vocabulary or intent. The illustrations and accompanying text are designed to help students gain a higher level of understanding. Students who complete this book will not only be conversationally literate, they'll also be able to pass their next bio test.

How This Book Is Organized

This book is presented in six sections:

In **Part 1, "Back to the Basics of Life,"** the requisite background chemistry information is provided in a detailed format that identifies and explains the function and structure of the biologically important chemicals. The next chapters establish the cell as the basic unit of function for all living things and identify common cellular functions and the structures responsible for the functions. They also explain some of the more specialized cellular functions and how they fit into the overall context of various organisms' lifestyles.

The chapters in **Part 2, "Genetics and Inheritance,"** present the popular science of heredity. Everything from the process of cell division to what happens when good cells go bad is presented with regard to plants, animals, and humans (as appropriate). Specific types of Mendelian and non-Mendelian inheritance are explained in detail. There are step-by-step instructions to help you solve those challenging inheritance problems.

In **Part 3, "Evolution, Natural Selection, and Speciation,"** the wonders of genetics are integrated into the reality of life on Earth to create the backdrop for Darwin's theories. The favoritism of natural selection is explained in terms of population dynamics, the Hardy-Weinberg equilibrium model, and speciation. The geologic time scale is overlaid to provide a time continuum that establishes an overall perspective that distinguishes micro- from macroevolution concepts. The battle over which classification system to use is examined, and the possible origin of all life that goes into the taxonomic system is examined from a number of different historical and scientific viewpoints. Theories that explain the origin of prokaryotes, modification to eukaryotes, and production of protists round out this comprehensive look at the pressures of the environment expressed in living things and the confusion that surrounds them.

Plants are in vogue in **Part 4, "Plant Diversity and Systems Analysis,"** as topics related to the function of plants are matched with specific structures that are cleverly designed to support that particular function. Plants are living organisms that have specific nutrient and environmental requirements and have interesting hormones that make them do funny things. The simplest plants give way to more complicated life cycles as the plant kingdom unfolds in a systematic fashion.

Part 5, "Animal Diversity," provides a detailed analysis of all the systems that affect you and your animal cousins. It also explains why animals exhibit nonrational behaviors that place them at risk. The animal kingdom is approached in a systematic way, from simple to complex, with humans placed at the top. You will learn the difference between a flatworm and a roundworm and why it is important to know the difference! You will also find out whether there are hominids living among us.

Finally, in **Part 6, "Biosphere and Ecology,"** all previous chapters come together as the biotic and abiotic factors are examined in a way that justifies the vegetative and animal life that live in a particular area and why they are still there, or not. The elements of population growth and dynamics are explored and related to the Malthusian dilemma and world order. The various types of conservation are explored in a manner that demonstrates how the environment is supposed to work, how it is currently working, and what we can do about it.

Things to Help You Out Along the Way

Learning and understanding build on prior learning and understanding and draw upon the background of the student to construct meaning. I have included useful items that may help establish linkages to other memorable material. If you are like me, it is important to establish a colorful background so that new information "fits" into those things that you already know.

If a new piece of information is added to my knowledge bank, it either connects with something already present or opens a new file. New files tend to deteriorate quickly over time. Connecting new learning with existing learning guarantees retrieval of the combined learning at a later time. Check out the extra help that has been provided for your illumination and enjoyment.

Bionote

Bionotes are related ideas that are designed to provide a memory aid, complete a related thought, or add peripherally to your knowledge bank. They are presented in a format that will accentuate understanding and help develop memory linkages.

Biohazard

A biohazard is a caution to all readers that something unhealthy may happen if you pursue a particular action. They are designed with the reader's health in mind. Usually they remind the reader of the reason why Mom told us not to do something.

Bioterms

Bioterms are definitions of specialized words that may be critical to the understanding of the text. Biology is like learning a foreign language; lots of terms are unfamiliar to most people.

Acknowledgments

I would like to acknowledge the help provided by my Heavenly Father in preparing me to write this text. No earthly power could have done it!

A few other people have been influential in allowing me the benefit of their friendship. It would be difficult to do anything, even something as simple as writing a book, without their help. I am thankful for the help of the following friends, listed in no particular order: Carol, Dave, Rick, Ron, Keith, Scot, Gail, Mike, Jim, Steve, Tim, John, Joe, Kerry, Tom, Bill, Panda, Mark, and X.

Part 1

Back to the Basics of Life

Although most people view biology as a study of plants and animals, it requires a fundamental background in chemistry and chemical compounds to really make sense of more complex topics. Each biomolecule will be revisited many times throughout the book. Please digest them slowly.

The structures and related functions of the fundamental unit of life, the cell, are presented so they may be built upon in later chapters. Don't panic because we are getting into biology; topics in this section are broken down into bite-size pieces.

Chapter 1

The Chemistry of Biology

In This Chapter

- Why you need to understand atomic theory
- Chemical reactions and energy change
- The uniqueness of water
- The importance of organic chemistry
- Carbohydrates, lipids, and proteins
- Nucleic acids, vitamins, and minerals

In the beginning, there was biology. It reigned supreme over the other subject areas in splendor and elegance. It defined and explained the living parts of the world and their environment. What else is there to know?

An understanding of biology requires a little knowledge of chemistry, and an understanding of chemistry requires a little knowledge of mathematics—that's where we draw the line. It is important to know aspects of chemistry to make biology come alive, but it is not important to go into detailed mathematical applications to understand the principles of biology—so we won't!

Modern biology overlaps with chemistry in explaining the structure and function of all cellular processes at the molecular level. Several important chemical concepts are treated in detail in the sections that follow. When applied in later chapters, these chemical concepts will allow you to construct greater meaning of the more complex biological principles.

Atomic Theory

For biological purposes, the simplest form of a pure substance that retains all of the properties of that substance is an *atom.* Atoms are also the simplest form of an *element.* An element is made up of only one type of atom. For instance, carbon is an element. All carbon atoms look the same and they are the only atoms that are carbon atoms. A different atomic structure would be a different atom and therefore a different element. Consult a periodic table for the complete listing of the 100+ elements (for example, pearl1.lanl.gov/periodic/default.htm). All atoms have three components: protons, neutrons, and electrons. The first element, hydrogen, has one proton, usually one neutron, and one electron. (Hydrogen exists in nature as a combination of three isotopes each of which has a different number of neutrons. Normally the most prominent isotope which has one proton is considered.) All hydrogen atoms for the most part look the same.

Protons are a positive charge and along with neutrons, which carry no charge (neutral), are found in the centrally located *nucleus* of the atom. Electrons are negatively charged and orbit around the nucleus in a three-dimensional cloud of energy levels. The positive-negative attraction (magnetism) keeps the electrons from flying away from the nucleus. Conversely, the *kinetic energy* (energy associated with motion) of the electron provides it with the energy of movement and keeps the electron from slamming into the nucleus. The balance between these two forces, magnetism and kinetic energy, defines the orbital pathways of all electrons. If a neutral atom loses or gains one electron (or more), such as might happen during a chemical reaction, the resulting atom has an imbalance between the existing charges. This imbalance creates either an excess positive (+) charge, if the atom lost an electron, or an excess negative (–) charge, if the atom gained an electron. Atoms that have a charge are called *ions,* and they are very reactive and fundamentally important in chemical reactions.

Chemical Reactions: Ionic, Covalent, and Polar Covalent Bonds

Chemical reactions are important to all levels of biology. In the simplest terms, a reaction requires reactants and products. Reactants are the atoms or molecules that are involved with the change, and products are the resulting changed atoms or molecules. In most biological reactions, *enzymes* act as *catalysts* to increase the rate of a reaction. A chemical reaction occurs when reactants are joined together to create a product that has different chemical properties than the original reactants. This always involves an energy change and a change in the electron configuration around the original atoms. When electrons redistribute their orbitals to include two or more atomic nuclei, as is the case in a *covalent* bond, or donate or accept electrons, as is the case in an *ionic* bond, a chemical reaction has occurred. Two general types of bonds form during chemical reactions: ionic and covalent.

Ionic bonds form when the outermost, or valence, electrons of an atom are donated or received in association with a second atom. Because the electrons are now orbiting around the receiving atom and not their original atom, the receiving atom now has an imbalance between the number of protons and electrons and becomes a negatively charged ion. The donating atom also has a proton-electron imbalance and becomes a positively charged ion because it lost a negatively charged electron and the number of its protons remained the same. The resulting molecule has properties different from the original atoms. It is important to remember that because of the unequal electron distribution around the reacting atoms, the resulting ionic compounds have partial charges. This importance is developed in greater detail in Chapter 3, but it explains the fact that water can dissolve any substance that has a partial charge on it. A typical example for an ionic bond is the joining of a sodium atom, which donates an electron, to a chlorine atom, which accepts the electron, to form sodium chloride, also known as table salt.

Covalent bonds occur when two or more atoms share their electrons. The electrons are not donated/accepted; instead, they incorporate their orbitals to create an electron cloud around all participating atoms. When the electrons are shared evenly around all reacting nuclei, there is no partial charge on the resulting molecule, as is the case when carbon covalently bonds with itself. However, in some cases, the electrons are not shared evenly and partial charges occur, as in the case of polar covalent bonds.

In reality, many bonds are actually a hybridization of ionic and covalent and have characteristics of both types. Atoms with *polar covalent* bonds share their electrons (covalent characteristic) unevenly (ionic characteristic), giving a slight positive (+) charge to one end of the molecule and a slight negative (–) charge to the other end. Water is a polar covalent molecule because the electrons spend more of their time around the oxygen atom because the oxygen atom has more protons acting as electron-magnets. Because of this uneven sharing of electrons, the oxygen end of the molecule has a slight negative charge, and the hydrogen end has a partial positive charge because the electrons are spending more time orbiting around the oxygen atom. The overall molecule has a partial positive and a partial negative end. As a result, water molecules tend to align themselves so that the positive end of one molecule aligns with the negative end of another molecule (opposites attract).

Notice also in the ionic model that the electrons are drawn away from the sending atom and accepted by the receiving atom. The covalent model shows the electrons being shared equally around all of the atoms, whereas the polar covalent shows the unequal sharing of the electrons.

In all cases, the driving force for any chemical reaction is a move toward greater stability of the atoms. To increase stability, atoms tend to react so that they lower their energy and increase their *entropy* (randomness or lack of organization). In chemical terms, this means that they seek to have a stable number of electrons in their outermost orbital. The stable number means that the outermost energy level is either completely full or completely empty. Chemists call this the *v* because often eight valence electrons are required to reach stability. Atoms react to achieve this electron configuration by donating/accepting electrons (ionic) or sharing them (covalent). Biomolecules are considered organic because they contain the element carbon and are covalently bonded.

Water

Water is one of the most unique molecules known to man and also one of the most important to biological systems. Not only does water exist in nature in all three states of matter (solid, liquid, gas), it also covers 75 percent of the earth and composes roughly 78 percent of the human body.

The uniqueness of water comes from its molecular structure. Because it is a polar covalent molecule, it has a slight positive and slight negative charge on opposite ends. Examine the illustration *Water molecule* and note two important characteristics. First, notice the location of the slight positive and negative ends. Second, observe that water is a bent molecule, not linear or straight.

Water molecule.

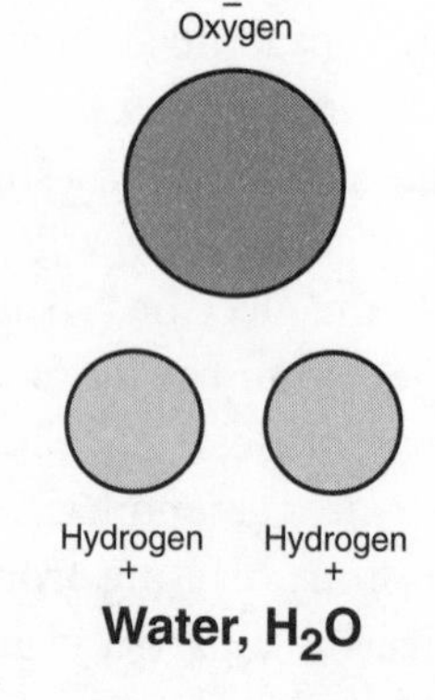

Because water is a bent, partially polar molecule, it possesses the following biologically important characteristics of what is formed by the joining of many water molecules—all of them are critical to the creation and support of life on Earth:

- Polarity
- Hydrogen bonding
- Cohesion
- Surface tension

Polarity

Polarity simply means that the molecule has both a positively and negatively charged end. More important, the polarity of water is responsible for effectively dissolving other polar molecules, such as sugars and ionic compounds such as salt. Ionic compounds dissolve in water to form ions. This is important to remember because for most biological reactions to occur, the reactants must be dissolved in water. Because water is able to dissolve so many common substances, it is known as the *universal solvent.* Substances that cannot be dissolved by water (such as oils) are called fat soluble and are nonpolar, nonionic compounds that are strongly covalently bonded. Insoluble substances make excellent containers of water, such as cell membranes and cell walls.

Hydrogen Bonding

When water molecules align with each other, a weak bond is established between the negatively charged oxygen atom of one water molecule and the positively charged hydrogen atoms of a neighboring water molecule. The weak bond that often forms between hydrogen atoms and neighboring atoms is the *hydrogen bond.* Hydrogen bonds are very common in living organisms; for example, hydrogen bonds form between the bases of DNA to help hold the DNA chain together. Hydrogen bonds give water molecules two additional characteristics: cohesion and surface tension.

Cohesion

Because of the extensive hydrogen bonding in water, the molecules tend to stick to each other in a regular pattern. This phenomenon, called *cohesion*, is easily observed as you carefully overfill a glass with water and observe the water molecules holding together above the rim until gravity overtakes the hydrogen bonding and the water molecules spill down the side of the glass. Likewise, the cohesive property of water allows tall trees to bring water to their highest leaves from sources below ground; please reference "Maintaining Water Balance" in Chapter 12.

Bionote

To avoid the pain of a belly-flop, divers enter the water with their hands or feet first to break the surface tension of water.

Surface Tension

A special type of cohesion is *surface tension.* The tension on the surface of water occurs when water molecules on the outside of the system align and are held together by hydrogen bonding to create an effect similar to a net made of atoms. For example, the surface tension of water allows water spiders to literally walk on water.

Organic Chemistry

Carbon is an especially noteworthy element in living systems. The chemistry of carbon, organic chemistry, is a complete study unto itself. We will touch on several highlights that will be useful in succeeding chapters.

The Uniqueness of Carbon

The carbon atom has four valence (outermost) electrons. Because of this unique configuration, it is easier for the carbon atom to share its four electrons with another atom or atoms than to lose or gain four electrons. Because each carbon is identical, they all have four valence electrons, so they can easily bond with other carbon atoms to form long chains or rings. In fact, a carbon atom can bond with another carbon atom two or three times to make double and triple covalent bonds between two carbon atoms. Long chains of carbon atoms with double and triple bonds are quite common in biology.

Carbon's tendency toward covalent bonding with itself generates three unique characteristics that create a vast array of compounds, including those necessary to construct and support life:

Bionote

Carbon to carbon (C-C) bonds form the backbone of all biomolecules and can include thousands of C-C bonds.

- The single bond that connects carbon atoms to carbon atoms is quite strong, so the subsequent long chains and ring structures are not fragile.
- The carbon-carbon covalent bonding pattern satisfies the Octet rule, making carbon compounds unwilling to react.

- Because carbon has four valence electrons and needs eight to satisfy the Octet rule, it can bond with up to four additional atoms, creating countless compound possibilities.

Functional Groups

In the simplest terms, the reactive part of any compound is called the *functional group*. Normally a functional group is a collection of atoms that operates as one reactive unit and is also the part of the molecule involved in a chemical reaction. Whereas carbon-to-carbon bonds are nonreactive, the instability of the functional groups drives chemical reactions that involve stable carbon-based compounds. For simplicity and reference, three functional groups are presented:

- *Amine* is identified by a central nitrogen atom that has three bonds, usually to hydrogen atoms. Amine groups form the basis for amino acids, which when bonded together form *proteins*.
- A *Carboxylic group (COOH)* is attached to the long carbon chains that form *fatty acid* molecules, which are a type of lipid *lipids*.
- *Hydroxyl groups (OH)* are very reactive. They are a component of alcohols, such as ethanol, the alcoholic component of adult beverages. The oxygen-hydrogen association is unique to this functional group and easily identifies it as an alcohol.

Dehydration Synthesis and Polymer Formation

Polymers are small molecules that can be bonded together to create larger molecules. Complex *carbohydrates* are made from small simple sugars joined together, and giant protein molecules are simply a series of smaller amino acid molecules bonded together. The prefix *poly* identifies this type of molecular addition. For instance, polysaccharides are large carbohydrates composed of multiple saccharide (sugar) units.

The chemical reaction that powers polymer formation is known by several names, including *dehydration synthesis* and condensation reaction. Regardless of the name, the molecules are joined by bonding sites created when a positively charged hydrogen ion (H^+) is lost from one molecule and a negatively charged hydroxide (OH^-) ion is lost from a neighboring molecule. The H^+

Bionote

A hydration reaction is the reverse of a dehydration reaction in that water serves as a reactant to split apart large poly-molecules.

and OH^- combine to form water. So a dehydration synthesis joins two smaller units together with the loss of one water molecule.

Carbohydrates

Carbohydrates are organic compounds that are organized as ring structures and are always composed of the elements carbon, hydrogen, and oxygen. Carbohydrates are truly hydrates of carbon because the ratio of hydrogen atoms to oxygen atoms is always nearly 2:1, as in H_2O.

They also have many functions. Most of the energy you receive comes from the carbohydrates that you eat. Plants make carbohydrates such as wheat, corn, and potatoes. Carbohydrates are normally consumed by animals either by eating the plant that manufactured it or by eating other animals. Humans also receive carbohydrates from whole grains, fruits, vegetables, milk, candy, soft drinks, and pasta.

Insects manufacture the carbohydrate chitin as a tough exoskeleton for protection, and lobsters and crabs use chitin for their shells. Finally, cellulose is probably the most widely used carbohydrate compound, comprising wood and wood products, such as paper.

Monosaccharides

The simplest biologically important carbohydrates are *monosaccharides*, meaning one sugar (mono = one, saccharide = sugar). The general formula for any carbohydrate is $(CH_2O)_x$ where x is any number between three and eight. The most common monosaccharides (hexoses) are glucose, galactose, and fructose.

Glucose is the simplest monosaccharide and probably the most familiar sugar, especially if you have been in the hospital. In nature, glucose is the sugar that green plants produce during *photosynthesis*. It is also the main source of energy for cells. Medical procedures often require a glucose IV for recovering patients to regain their strength more quickly. Galactose is found in milk, and fructose gives fruit a sweet flavor. Although the chemical structure of each sugar differs, the chemical formula is the same: $C_6H_{12}O_6$.

Disaccharides

Monosaccharides are joined together by *dehydration synthesis* to form disaccharides, or double sugars (di = two). The dehydration synthesis reaction releases water (dehydration) as a by-product. The most common disaccharide is sucrose, also known as table sugar, $C_{12}H_{22}O_{11}$. Other common disaccharides include maltose (malt sugar) and lactose (milk sugar).

Polysaccharides

Further dehydration adds more sugar molecules together to form long chains known as *polysaccharides*. A polysaccharide generally refers to a carbohydrate polymer consisting of hundreds, even thousands of monosaccharides covalently bonded together. Cells use polysaccharides for a number of reasons, including the storage of excess glucose as starch in plants and glycogen in animals. The large polysaccharide cellulose is a structural component found in plants that gives them their rigidity and flexibility.

Lipids

Lipids are organic compounds that contain the same elements as carbohydrates: carbon, hydrogen, and oxygen. However, the hydrogen-to-oxygen ratio is always greater than 2:1. More important for biological systems, the carbon-to-hydrogen bonds are nonpolar covalent, which means that lipids are fat soluble and will not dissolve in water. There are four biologically important lipids:

- Fats
- Waxes
- Phospholipids
- Steroids

Fats

Fats are large molecules that are composed of three *fatty acid* molecules bonded to a *glycerol* molecule. The fatty acid molecule is a long chain of covalently bonded carbon atoms with nonpolar bonds to hydrogen atoms all along the carbon chain with a carboxyl group attached to one end. Because the carbon-hydrogen bonds are nonpolar, the chain is *hydrophobic,* meaning they are not water soluble. Glycerol is a three-carbon-chain compound that bonds with the fatty acids to create a fat. Typically, each carbon in the glycerol molecule bonds via dehydration synthesis to the first carbon atom from a fatty acid molecule so that the resulting fat molecule appears to have a glycerol head with three fatty acid chains streaming from it. This resulting molecule is called a *triglyceride*. Because carbon-hydrogen bonds are considered energy rich, fats store a lot of energy per unit. In fact, a gram of fat stores more than twice as much energy as a gram of a polysaccharide such as starch. Fats are lipids that are used by living organisms for stored energy.

A saturated fatty acid has hydrogen atoms bonded to all available carbon atoms. An unsaturated fatty acid has one or more carbon atoms double-bonded to the neighboring carbon atom so that fewer hydrogen atoms are needed to create a stable electron cloud. With fewer hydrogen atoms attached, the molecule is considered unsaturated with hydrogen atoms. So saturated fatty acids have more hydrogen atoms attached than unsaturated fatty acid chains. Through their metabolism, plants generally produce triglycerides that contain unsaturated fatty acids such as peanut oil or olive oil, whereas animals generally produce triglycerides that contain saturated fatty acids which humans sometimes convert into butter and lard.

Waxes

Waxes are similar to fats except that waxes are composed of only one long-chain fatty acid bonded to a long-chain alcohol group attached. Because of their long, nonpolar carbon chains, waxes are extremely hydrophobic (meaning they lack an affinity for water). Both plants and animals use this waterproofing characteristic as part of their composition. Plants most noticeably use waxes for a thin protective covering of stems and leaves to prevent water loss. Similarly, animals employ waxes for protective purposes; for instance, earwax in humans prevents foreign material from entering and possibly injuring the ear canal area.

Phospholipids

Phospholipids are similar to fats except they have two fatty acid chains bonded to a glycerol plus they contain the element phosphorus. Phospholipids are unique because they have a hydrophobic and a *hydrophilic* (water-soluble) end. Phospholipids are biologically important because they are the main structural components of cell membranes. The cell membrane is called a phospholipid bilayer because it consists of two phospholipid layers oriented so that the hydrophyllic "head" of both molecules face the exterior and the hydrophobic "tails" of both molecules create the interior of the membrane. Therefore, water and other cellular fluids are contained. The hydrophobic ends for both molecules face each other on the inside and allow for passage of acceptable, and some objectionable, materials through the cell membrane.

Steroids

Steroids are structurally different from the other lipids. The carbon skeleton of steroids is bent to form four fused rings that do not contain fatty acids. The most common steroid, cholesterol, is needed to make both the male (testosterone) and female (estrogen) sex hormones, and it is a component of cell membranes and is needed for the proper function of nerve cells. Excessive amounts of cholesterol, however, have been linked to heart disease. Another popular steroid group is the anabolic steroids that are man-made and mimic the effect of the male hormone, testosterone. Originally intended as a treatment for anemia and certain diseases that destroy muscle, athletes have recently been using them to increase muscle mass, stamina, and strength—which they will do. However, the performance-enhancement drugs come with a price. The anabolic steroids are linked to increased cholesterol levels, mood swings, reduced sex drive, possible infertility, and possible connections between liver damage and the resulting liver cancer. Certain beneficial fat-soluble hormones, such as cortisol, are also familiar steroids.

Proteins

Proteins are organic compounds that contain the element nitrogen as well as carbon, hydrogen, and oxygen. Proteins are the most diverse group of biologically important substances and are often considered to be the central compound necessary for life. In fact, the translation from the Greek root word means "first place." Skin and muscles are composed of proteins; antibodies and enzymes are proteins; some hormones are proteins; and some proteins are involved with digestion, respiration, reproduction, and even normal vision, just to mention a few.

Amino Acids

There are obviously many types of proteins, but they are all made from *amino acids* bonded together by the dehydration synthesis. By continually adding amino acids, called peptides, two amino acids join together to form dipeptides; as more peptides join together, they form polypeptides. Proteins vary in length and complexity based on the number and type of amino acids that compose the chain. There are about 20 different amino acids, each with a different chemical structure and characteristics; for instance, some are polar, others are nonpolar. The final protein structure is dependent upon the amino acids that compose it. Protein function is directly related to

Bionote

Humans must obtain nine essential amino acids through their food because our bodies are not capable of manufacturing them. A missing amino acid restricts the protein synthesis and may lead to a protein deficiency, which is a serious type of malnutrition. Remedy: Eat lots of corn, grains, beans, and legumes as part of your normal, balanced diet.

the structure of that protein. A protein's specific shape determines its function. If the three-dimensional structure of the protein is altered because of a change in the structure of the amino acids, the protein becomes *denatured* and does not perform its function as expected.

Protein Structure

The three-dimensional geometry of a protein molecule is so important to its function that four levels of structure are used to describe a protein. The first level, or *primary structure*, is the linear sequence of amino acids that creates the peptide chain. In the *secondary structure*, hydrogen bonding between different amino acids creates a three-dimensional geometry like an *alpha helix* or *pleated sheet*. An alpha helix is simply a spiral or coiled molecule, whereas a pleated sheet looks like a ribbon with regular peaks and valleys as part of the fabric. The *tertiary structure* describes the overall shape of the protein. Most tertiary structures are either globular or fibrous. Generally, nonstructural proteins such as enzymes are globular, which means they look spherical. The enzyme amylase is a good example of a globular protein. Structural proteins are typically long and thin, and hence the name, fibrous. *Quaternary structures* describe the protein's appearance when a protein is composed of two or more polypeptide chains. Often the polypeptide chains will hydrogen bond with each other in unique patterns to create the desired protein configuration.

Enzymes

Most *enzymes* are proteins and therefore their function is specific to their structure. Enzymes function as a catalyst to increase the rate of virtually all the chemical reactions that take place in a living system. The enzymes, like all catalysts, are not consumed but are constantly reused to catalyze the same specific reaction. Enzymes depend on the correct structural alignment and orientation at the *active site* of the protein and the appropriate site of the reactants, or *substrate*, before the reaction can proceed. This geometric interaction between the enzyme and the substrate is referred to as the "lock-and-key model" because the enzyme's action parallels the action of a lock into which is fitted the key (substrate). If the key and lock do not match, the action does not work.

It is the same with enzymes and substrates. The active site for the enzyme and the appropriately matched site of the substrate must physically join before the reaction can occur. That is why the structure of the enzyme is so important. The enzyme binds with the appropriate substrate only in the correct alignment and orientation to connect the molecules. The resulting enzyme-substrate complex enables the reaction to occur.

Finally, the products are formed and the enzyme is released to catalyze the same reac-tion for another substrate of the same type of molecule. Enzymes may fail to function if they are denatured. Remember the model simplifies your understanding of the process; in reality they are three-dimensional molecules.

Hormones

Hormones are chemical messengers produced in one part of the body to function in a different part of the body. Although fat-soluble hormones are made from steroids, water-soluble hormones such as the growth hormone are made from amino acids. Hormones function similarly to enzymes in that both require a specific receptor and perform a specific function. After a hormone is created and secreted by a cell, it travels—usually via the bloodstream—to its *target cell*. The target cell is the point of action that the hormone recognizes, binds to, and thereby delivers the chemical mes-sage. The hormone identifies the target cell by its receptor protein and employs the same lock-and-key process.

Nucleic Acids

Nucleic acids, which are composed of nucleotides, are very large and complex organic molecules that contain the genetic code for that organism. Two closely related types are needed to transmit the genetic information from parent to offspring: DNA and RNA. Not surprisingly, they also share structural similarities.

Nucleotides

Both DNA (deoxyribonucleic acid) and RNA (ribonucleic acid) are polymers of individual *nucleotides*. Each nucleotide has three components:

- Five-carbon-ring sugar (deoxyribose or ribose)
- Phosphate group
- Nitrogen base

Both DNA and RNA have four nitrogen bases available to construct nucleotides. Three of the nitrogen bases are the same. Review the following table for similarities and differences between DNA and RNA.

Bases

	RNA
	Adenine
Thymine	Uracil
Cytosine	Cytosine
Guanine	Guanine

Both DNA and RNA are large molecules of successive nucleotides bonded to the main chain by the dehydration synthesis reaction. Additional nucleotides connect when the phosphate of one nucleotide bonds with the nitrogen base of the next nucleotide.

DNA

Francis Watson and James Crick discovered the structure of DNA, which today we refer to as a double helix. The DNA double helix is actually two complementary strands of DNA that wrap around each other and are held in place by hydrogen bonding between the two strands. The resulting structure is a spiraling geometric molecule that looks like a twisted ladder, a.k.a. a double helix. It is important to remember that DNA molecules are so large they cannot escape the *nuclear membrane* that surrounds the nucleus of a cell.

Bionote

During asexual reproduction, an exact copy of the parent DNA is copied and supplied to all offspring so that both parent and offspring are identical.

Each DNA molecule consists of many *genes*. Each gene is composed of hundreds or usually thousands of nucleotides arranged in a specific order. The unique sequencing of the nitrogenous bases associated with the nucleotides dictates the primary structure of the protein to be created by the cell. Your chromosomes and those of all living creatures consist of long strands of DNA that store the heredity information, or genetic code, for that organism.

RNA

RNA (ribonucleic acid) is similar to DNA in that both are composed of long chains of nucleotides. However, RNA is a single strand and therefore much smaller than DNA. Although DNA contains the genetic information that determines all cell characteristics and functions, RNA stores and transfers the genetic code contained in the DNA. Its smaller size allows RNA to move freely through the nuclear membrane to transport the genetic instructions, usually to the *ribosomes*, for protein assembly.

Vitamins and Minerals

Vitamins are organic compounds that are required by advanced animals in small amounts on a regular basis. Like essential amino acids, there are 13 vitamins that are required but not produced by the human system. Unlike essential amino acids, these vitamins are required in minute quantities. If deficient or in excess, certain maladies occur in humans, such as beriberi, anemia, rickets, and skin lesions. In general, vitamins are *coenzymes*, or parts of enzymes, that function to assist a specific enzyme to catalyze (increase the rate of) a reaction. Some vitamins are fat soluble and others are water soluble.

Fat-soluble vitamins are probably the most common vitamins for some people. Unlike water-soluble vitamins, they remain stored in the fat deposits of a body for long periods of time and may accumulate to overdose levels. Note that the only vitamin humans can make is Vitamin D. Vitamin D is made when cholesterol is acted upon by enzymes and sunlight. It should also be noted that the fat-soluble Vitamin K is produced in small quantities in the human intestine by the action of mutually beneficial intestinal bacteria.

Water-soluble vitamins generally function within the cell to help catalyze cellular reactions such as *cellular respiration*. For your reference, cellular respiration is the process of harvesting energy from the breakdown of food molecules that takes place inside individual cells. Unlike fat-soluble vitamins, excess water-soluble vitamins do not remain stored in the body, but are excreted in urine and feces. Water-soluble vitamins include the eight different types of B complex vitamins and Vitamin C.

Minerals are naturally occurring inorganic substances required in trace amounts for normal body functions such as the development of strong bones and teeth, proper muscle and nerve functions, and construction of red blood cells. Like vitamins, these essential minerals are not produced by humans, so they must be consumed on a regular basis. Because they are water soluble, excessive amounts are eliminated through normal urinary functions and perspiration.

The Least You Need to Know

- Chemical principles explain many biological processes.
- The unique structural and chemical properties of water support life.
- Carbon has unique chemical characteristics that make it a central element in the structure and function of all living things.

- Biomolecules compose all living things and determine their structure and function.
- The structure of a molecule determines its ability to function correctly.
- There are 13 vitamins required for proper human health, including Vitamin D, which is the only one that can be made by the human body.
- Minerals are inorganic substances that are needed in small amounts to help regulate body functions.

Chapter 2

Cell Theory, Form, and Function

In This Chapter

- The relationship between cell volume, surface area, overall size and productivity
- Structural features that characterize prokaryotes as simpler than eukaryotes
- Usefulness and universal use of cellular organelles
- How the cell membrane allows and inhibits certain substances
- How cells create more cells
- Why viruses are so hard to control

Deep inside you, on your surface, and all parts in between, fundamental functional units called cells are busy 24/7 keeping your body in a living condition. Curiously, we aren't really in charge of their behavior! In fact, if we were, it is likely we would be in a state of nonliving because of the numerous activities that take place in every cell at all times. Thankfully, we have a nervous system that handles that for us and does not bother us with the trivia of everyday functions. This is an example of the great and miraculous way your body is structurally and functionally composed to address the pressures of the living world.

Plants have cells very similar to yours. So do the other animals. Humans are classified as animals—gasp!—because our cells look and act remarkably like all the other animal cells. Take a look at one of your cells under a microscope, then compare it with a similar one from a duck-billed platypus, if you can find one. If you switched the cells and handed them to a friend, the friend would likely not be able to tell them apart! Try it with a plant cell. But don't bet any money on it this time. The plant cell is likely to have some green things in it that are a sure giveaway. It also has a cell wall, and you don't need one. There are some fundamentals that go into every cell, uniting the world of living things into a oneness of the universe; there are also cellular modifications that make some cells look like they are from a different planet. We can sort it all out in this chapter. Perhaps you will look at your dog or a plant through new eyes after reading this chapter.

Cell Theory

While observing dead cork samples with a crude lens, Robert Hooke identified and named "cells." He thought that the small, simple units looked like the bare prison cells of his time, and the name *cell* stuck. His work launched a new frontier in scientific exploration that led to modern cell theory:

- All living things are made of cells.
- Cells are the basic units of structure and function in all living things.
- All cells come from the reproduction of existing cells.

Size Limitations

Most plant cells are approximately 0.002 inches in diameter, whereas most bacteria are even smaller at 0.000008 inches long (10 to 50 nanometers in metric units), making them impossible to see without magnification. Cell size is limited due to the inability of very large cells to provide nutrients and water and remove wastes in an efficient manner. The size limitation is due to the ratio between their outer surface area and the internal volume, making large cubical or spherical cells too big for the surface areas to accommodate all of their cellular life functions. Cells are three-dimensional, so as the cell grows, the volume increases geometrically as the cube of the side length, but actual surface area increases arithmetically with the square of the side length. In other words, a cell's volume increases more rapidly than the surface area. This becomes biologically important when a cell becomes too large for the available surface area to allow passage of nutrients and oxygen into, and cellular waste out of, the cell. Conversely, smaller cells can move materials in and out through the cell membrane at a faster rate because they have a more favorable surface area-to-volume ratio. Interestingly, the shape of muscle and nerve cells tend to be long and thin, which also provides a favorable surface area-to-volume ratio.

Bionote

Nerve cells are often long and fibrous-looking. The nerve cell in the leg of a giraffe is often longer than six feet.

Prokaryotes and Eukaryotes

Two structurally distinct types of cells have evolved that vary greatly in their internal complexity. *Prokaryote* cells are the simplest type and are evolutionary precursors to *eukaryote* cell types. What is thought to be the earliest known fossilized cells were discovered by paleontologists working near the Great Lakes in North America. They discovered microfossil evidence with enough detail to classify the cells as prokaryote. How did they know they were prokaryote?

Although both prokaryote and eukaryote cells can have a *cell wall* and a *cell membrane* to enclose the cellular *cytoplasm*, the structural similarities end there. Inside a typical prokaryote cell, such as a bacteria cell, there are no membrane-bound *organelles*. An organelle is a subcellular structure that has a specific function. Even the genetic material, although often contained and cornered inside the cell, is not bound by a membrane. Eukaryotic cells, which basically include every cell type except bacteria, are characterized by internal organelles surrounded by a membrane, which helps to increase their organization and efficiency. In contrast to prokaryotes, in eukaryotes the chromosomes are made of distinct lengths of DNA and are stored within a nuclear membrane. Because prokaryotes are simpler, lacking membrane-bound organelles, they are also much smaller (1 to 10 micrometers) than eukaryotes, which range from 10 to 100 micrometers in size.

Viruses

Viruses are small nucleic acid units, either DNA or RNA, surrounded by a protective protein coat, or *capsid*, making them little more than packaged genes. Some viruses, such as influenza (flu), have a cloaking protein envelope, making it easier to penetrate a host cell. Other viruses such as HIV also have an unusual complement of enzymes that create interesting products. In general, their overall size ranges from 20 nm to 250 nm (one nanometer, nm = 0.00000004 inches), making them much smaller than any single-celled organism and only visible through an electron microscope. Viruses have plagued man for millennia, causing such human maladies as chickenpox, warts, hepatitis, smallpox, polio, mononucleosis, colds, herpes, and rabies, just to mention a few. Although viruses contain either DNA or RNA, they are actually considered non-living because they do not grow, reproduce on their own, maintain homeostasis, nor metabolize. Their "life cycle" is an interesting study of deception, pillage, and piracy.

Bioterms

A **bacteriophage** is a virus that infects bacteria. They are useful because they are easy to study and have added greatly to our knowledge of viruses and how they work.

A virus, also known as a phage, can only survive by infecting a living host cell and turning that cell into a factory to manufacture more viruses. Research in molecular biology often studies *bacteriophages* because they are common and easier to culture and maintain than more pathogenic types. There are two known methods which explain how viruses are spread: *lytic cycle* and *lysogenic cycle*.

Lytic Cycle

In the lytic cycle, the phage always destroys the host cell as the final act of the following five-part event:

1. The phage attaches to the cell membrane and injects viral DNA or RNA into the living host cell.
2. Injected phage nucleic acids contort into a circle inside the cell.
3. The infected cell mistakenly copies the phage DNA or RNA (whichever nucleic acid the phage possesses).
4. The copied nucleic acids organize as phages.
5. When the number of completely assembled phages becomes too large for the host cell to contain, the cell membrane breaks, releasing numerous phages to infect neighboring cells.

Lysogenic Cycle

The lysogenic cycle also has five stages, but the host cell is not destroyed, but is used to continually reproduce more phages:

1. The phage attaches to the cell membrane of a living cell and inserts its DNA or RNA.
2. Phage DNA or RNA reforms as a circle inside the host cell.
3. Phage DNA becomes incorporated into the host cell DNA, called a *prophage.*
4. Host cell reproduces normally and mistakenly makes new phage nucleic acids at the same time as normal nucleic acids. The phages are released into the environment to infect other cells.
5. Under certain conditions, a prophage may switch to a lytic pathway. Otherwise the host cell continues to generate more phages.

Certain viruses have a cloaking cover made of a protein-lipid combination with glycoprotein projections from the surface. These viruses, such as mumps, use their glycoprotein spikes to simulate a normal protein and thereby mask their identity so they can attach to receptor sites on the cell membrane of the host. The envelope then fuses with the cell membrane and allows the viral nucleic acids to spill into the host. Vaccines have been developed and are effective in combating certain viral diseases such as smallpox, mumps, and polio. A vaccine is a harmless variation of the microbe that is designed to stimulate the immune system of the individual. Chapter 15 presents a more detailed explanation of the human immune system.

Fluid Mosaic Model of Membrane Structure and Function

Membranes have many different functions within a typical cell, such as keeping unwanted viruses out, but probably the most valuable is the partitioning of the cell into functional and segregated compartments. Because of the incredible number and often conflicting biochemical reactions occurring in a cell at any one time, the cell must retain order via structural organization or risk chemical chaos. The internal membranes compartmentalize reactions to prevent interference. The cell membrane also separates life from the nonlife on its exterior. In so doing, an intact and healthy membrane is *selectively permeable* because it allows substances needed for cell prosperity to enter and attempts to prohibit the penetration of unwanted and unfriendly substances. Unfortunately the system is not always fool-proof. Sometimes unwanted substances pass through the membrane and may cause trouble within the cell.

Interestingly, when a phospholipid is placed in water, it spontaneously folds upon itself to create a double layer, or bilayer. This bilayer phenomenon is also the foundation for the widely upheld fluid mosaic model of membrane structure. The phospholipid molecule has a water-soluble, polar "head" and two fat-soluble, nonpolar "tails." The hydrophobic tails always try to avoid water and face the inside of the bilayer, whereas the hydrophilic head faces the exterior and the interior.

Bioterms

A **glycoprotein** is a molecule used as an identification or address for proteins seeking a particular site for bonding. There are many different types of glycoproteins because of the vast array of sugars that may combine with the proteins that compose them.

Within the phospholipid bilayer are many different types of embedded proteins and cholesterol molecules whose presence spawned the term *mosaic.* From scanning electron microscope images, it was observed that the embedded molecules can move sideways throughout the membrane, meaning the membrane is not solid, but more like a *fluid.* The membranes also have *glycoproteins* attached to their surface, which aid in their location and identification of food, water, waste, and other membrane traffic. Each cell has a particular glycoprotein structure based on its need to attract or repel membrane traffic. Refer to the illustration *Typical membrane,* and note the arrangement of the phospholipid molecules.

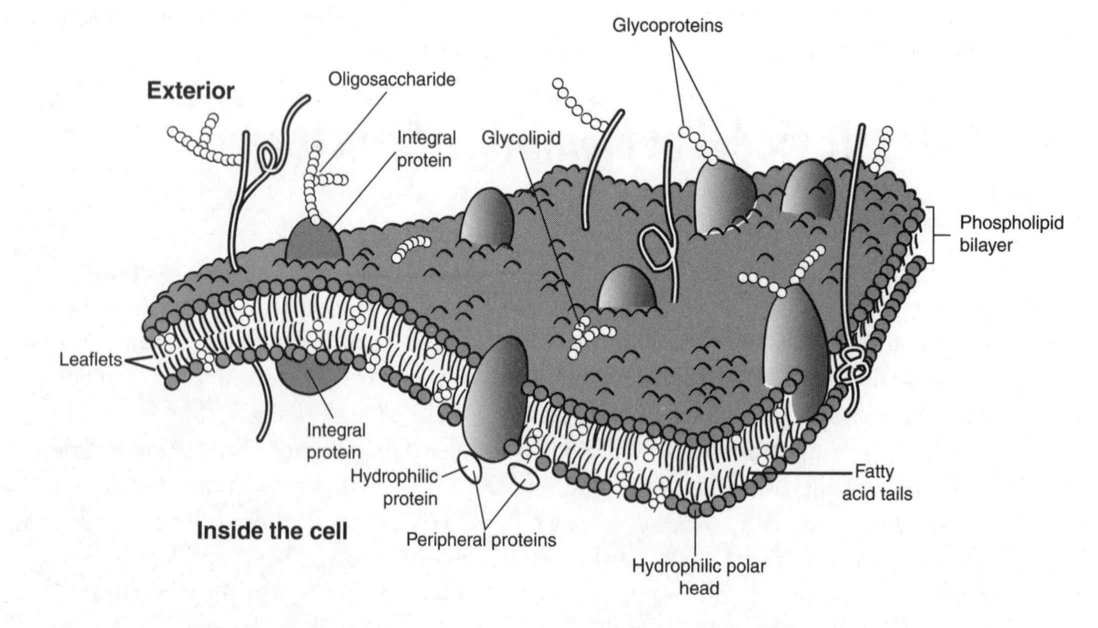

Typical membrane.

The proteins embedded in the membrane serve many of the membrane functions, such as holding the membrane in a regular, identifiable structure for easy bonding. They also have a specific and unique shape that allows them to function as *receptors* and receptor sites for attachment to the appropriate raw materials needed for cellular functions. In some cases, the receptor protein is also a *signal transducer* that begins a series of enzyme-catalyzed reactions to stimulate a particular reaction or function within a cell. Finally, the *transport proteins*, also called *carrier proteins*, help substances move across membranes, as described in the next section.

Passive Transport

Passive transport occurs when no energy is required to move a substance, such as water or carbon dioxide, from an area of high concentration to an area of low con-centration until the concentration is equal, sometimes across a membrane. The high-to-low concentration gradient is the driving force for passive transport because it fulfills a fundamental law of nature: Things tend to move from a high-energy, ordered structure to a lower-energy, increasing randomness, or increasing entropy state of being. The following are the classes of passive transport:

- *Diffusion*. This is a good example of how certain molecules, such as oxygen, simply move directly through a membrane in response to the high-to-low concentration gradient. As an example, oxygen diffuses out of the lungs and into the blood for transport to all of the cells.
- *Facilitated Diffusion*. This is a special type of diffusion that is useful because substances are sometimes too large to move freely through a membrane, or they need to move against a concentration gradient so transport proteins embedded in the membrane assist with the passage. In most cases, the transport protein creates a chemical channel for the passage of a specific substance. Because no energy is expended, the rate of facilitated diffusion depends on the number of transport proteins embedded in the membrane. As an example, glucose is moved by a glucose-transporter protein as it passes through the red blood cell into a body cell.
- *Osmosis*. This is similar to diffusion except that it refers only to water diffusing through a permeable membrane. Water as a solvent moves from an area of high to low concentration. In biological systems, it is easier to think of water as flowing from a low-solute to a high-solute concentration until the concentration is equal. The solution that has a high-solute concentration is a *hypotonic* solution relative to another lower-solute concentration or *hypertonic* solution. Water will continue to osmotically move from the low-solute/high-solvent concentration toward the high-solute/low-solvent concentration until both sides are *isotonic*, or equal. Osmoregulation is a struggle for all organisms as we continually adjust our cellular water balance for optimal conditions. In your body, the large intestine reabsorbs water by osmosis to help maintain the proper water concentration, which helps to keep your systems from dehydrating.

- *Ion channels*. These are membrane proteins that allow the passage of ions that would ordinarily be stopped by the lipid bilayer of the membrane. These small passageways are specific for one type of ion, such that a calcium ion could not pass through an iron ion channel. The ion channels also serve as gates because they regulate ion flow in response to two environmental factors: chemical or electrical signals from the cells and membrane movement. This happens in your body when a nervous impulse encounters a gap or synaptic cleft between nerve cells. The electrical stimulation is continued because ion channels are opened to allow specific ions to pass through the receiving membrane, which continues the electrical stimulation to the next nerve cell.

Active Transport

Sometimes substances must be pumped against a concentration gradient, such as the sodium ions (Na^+) and potassium ions (K^+) pump. So a transport protein and energy, usually adenosine triphosphate (ATP), the energy-rich compound, are needed to push the ions against the gradient. In the case of sodium and potassium ions, maintaining sodium outside and potassium inside the cell is crucial to the functioning of muscles and nerves. The following mechanism illustrates an *active transport* mechanism:

1. Sodium ions inside the cell bind to the transport protein as a phosphate is added from an ATP, which changes the shape of the transport protein.
2. The new transport protein structure carries and deposits the sodium to the exterior and bonds with a potassium ion, loses the phosphate group (which again changes the shape of the transport protein), and allows for the return trip.
3. The potassium is deposited inside the cell, and a sodium ion and a phosphate are attached to a transport protein to repeat the process.

Endocytosis and *exocytosis* handle the really big molecules, such as long protein chains or ringed structures, as well as the bulk volume of small molecules. In endocytosis, substances such as food are brought into the cell in a process in which the cell membrane surrounds the particle and moves the particle inside the cell, creating a vacuole or vesicle as a membrane-enclosed container. In exocytosis, waste products or hormones, which are contained in vacuoles or vesicles, exit the cell and their containing membrane is absorbed and added to the cell membrane. There are three types of endocytosis:

- *Pinocytosis* occurs when the cell absorbs fluid from the exterior, creating a fluid vacuole.

- *Receptor-mediated endocytosis* is a special type of pinocytosis that
identification of a receptor protein sensitive to the specific substa
- *Phagocytosis* is the engulfing and digesting of substances, usually foo
uoles with a lysosome attached (a lysosome is an organelle that conta
tive enzymes).

Endomembrane System

Unlike a prokaryote cell, all eukaryotic cells, regardless of plant, animal, or other origins, are structurally similar and contain mostly the same organelles, with certain exceptions noted. Eukaryotes are compartmentalized by inner membranes to increase active surface area, increase the sophistication of subcellular reactions, and thereby increase overall efficiency.

Within the eukaryotic cell, the endomembrane system is a functional association of membrane-bound organelles that are interconnected or closely connected that build, store, and transfer biomolecules. The biologically important endomembrane organelles are discussed in greater detail in the sections that follow. Refer to the illustration *A typical cell.*

The *nucleus* is the centerpiece of the cell. It stores the DNA in the form of *chromatin*, which is DNA plus a protein, and also serves as the site where RNA copies DNA to begin protein synthesis. The proteins are made in the ribosomes, which are in turn made by the nucleolus, which is also a nuclear component. The nucleolus is where the ribosomal RNA is made and assembled with proteins to make tribosomal subunits. A double *nuclear membrane* encompasses and separates the nucleus from the cytoplasm. Prokaryotes do not have a well-defined nucleus.

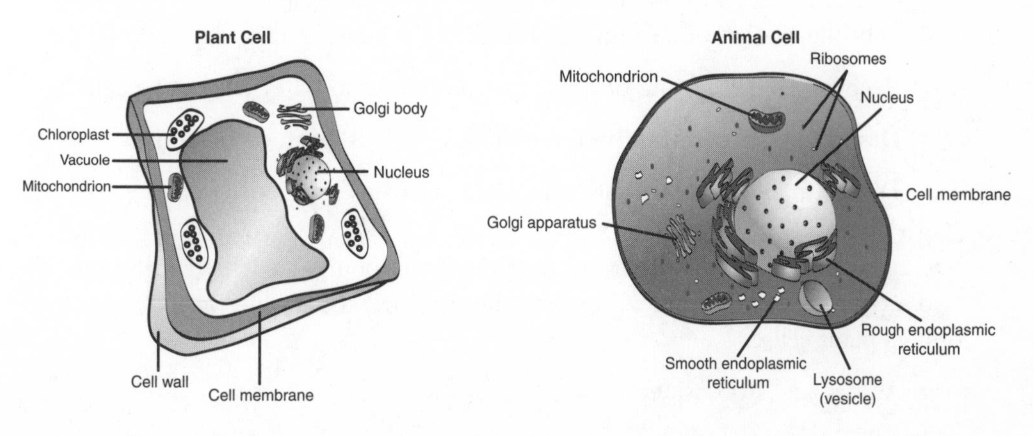

A typical cell.

ic reticulum is a long, continuous membrane that has
e rough endoplasmic reticulum (rER) is prominent in
s because it has ribosomes attached to the membrane.
ns, including making more membrane to lengthen
the cell. Another function is to make *secretory*
ated to function outside of the mother cell. Se-
lar exports are enveloped by a *transport vesicle* and
isturbed through the cytoplasm to the *Golgi apparatus.*

the nuclear membrane to the smooth ER (sER). The sER builds
e calcium levels so muscles perform correctly, and helps break down
nces in the liver. They do not contain ribosomes.

e Golgi apparatus, or Golgi complex, is not connected to the ER, but receives proteins from the ER and modifies them for extracellular export. When a Golgi apparatus receives a transport vesicle from the ER, it sorts the product into like storage areas, chemically marks them for destination points, repackages them in a new transport vesicle, and releases the resulting package to the cell membrane for extracellular export.

The structure of the Golgi apparatus supports its function. Electron microscopy indicates that the structure resembles interconnected, deflated balls or bags. One end serves as a "receiving point," the other a "shipping center," and the middle contains molecules that sort products and synthesize vesicles to surround them.

Lysosomes are sometimes called the cell's suicide pill because they are single-membrane organelles that contain hydrolytic, digestive enzymes that could easily destroy the cell. Their construction begins in the rER where the enzymes and membranes are joined, and finish in the Golgi apparatus. When fully functional, lysosomes are released and operate mostly in animal cells to perform their four primary functions:

- Subcellular digestion of food particles and nonfunctioning organelles
- Recovery and recycling of certain biomolecules for later use by the cell
- Destruction of harmful foreign particles, such as invading bacteria
- Digestion and removal of the webbing between embryonic fingers

Vacuoles, such as lysosomes, are single-membrane-bound sacs filled with fluid. They also serve four major functions, as explained by their use in the *central vacuole* found only in plant cells:

- Absorb and store water
- Store enzymes until needed, and metabolic wastes until removed

- Contain attractive pigments to lure pollinators to flowers
- Store toxic chemicals, which also serve as deterrents to herbivores

Energy Production: Chloroplasts and Mitochondria

All green plants have chloroplasts that serve as the location for photosynthesis. Although chloroplasts may be found in all above-ground parts of the plant, most are concentrated in the middle, or *mesophyll*, of the leaf. Chloroplasts are enclosed in a double membrane that creates a fluid-filled compartment between the membranes, called a *stroma*. Within the stroma are *thylakoids*, which are stacked like chips into *grana*. Within the thylakoid membrane are various types of chlorophyll molecules that capture and convert the energy of light into the chemical energy of chemical bonds. The thylakoid membrane greatly increases the available surface area and houses most of the enzymes and machinery for use in photosynthesis. Each photosynthetic cell contains many chloroplasts, which contain many grana.

Like all subcellular organelles, the function of the mitochondria is related to its structure. The primary purpose of the mitochondria is to conduct cellular respiration, converting the chemical energy of food molecules, such as carbohydrates, into high-energy compounds, such as ATP. Similar to chloroplasts, mitochondria are enclosed by a double membrane that creates a fluid-filled *intermembrane space*. The second compartment, the *mitochondrial matrix*, is contained by the highly folded inner membrane. The *cristae*, or folds of the inner membrane, greatly increase the surface area and contain a multitude of enzymes, so most cellular respiration reactions that produce ATP are completed in the mitochondrial matrix.

Endosymbiosis

An American researcher, Lynn Margulis, proposed in 1966 the hypothesis of *endosymbiosis*, which may explain the advent of the first eukaryote. According to Margulis, there were two successful invasions of an early anaerobic (one not requiring oxygen) prokaryote, by smaller independent prokaryotes. One of these prokaryote invaders entered the larger prokaryote probably for protection and easy access to nutrients, decided to stay, and began to reproduce independently inside the host cell. Rather than try to evict the invader, the two cells developed a mutually beneficial relationship. The invading cell is thought to be the modern-day mitochondria. A second invasion of similar style, but this time by a photosynthetic bacterium, eventually became a chloroplast. Interesting evidence supports this hypothesis. First, both mitochondria and chloroplasts contain their own DNA, which is separate and different from the rest of the cell. Second, the arrangement of their DNA is circular, a characteristic of prokaryote cells. Finally, both reproduce independently of the rest of the cell.

Cell Cycle: Interphase, Mitosis, Cytokinesis

Each eukaryotic cell has a repeating set of events that make up the life of every cell, called the *cell cycle.* Although they vary in length depending upon the cell's function, the cell cycle for all cells can be described in five steps. The first three steps where the cell grows, matures, and carries out its life function are collectively called *interphase,* followed by *mitosis,* and *cytokinesis.* Refer to the illustration *Cell cycle.*

Cell cycle.

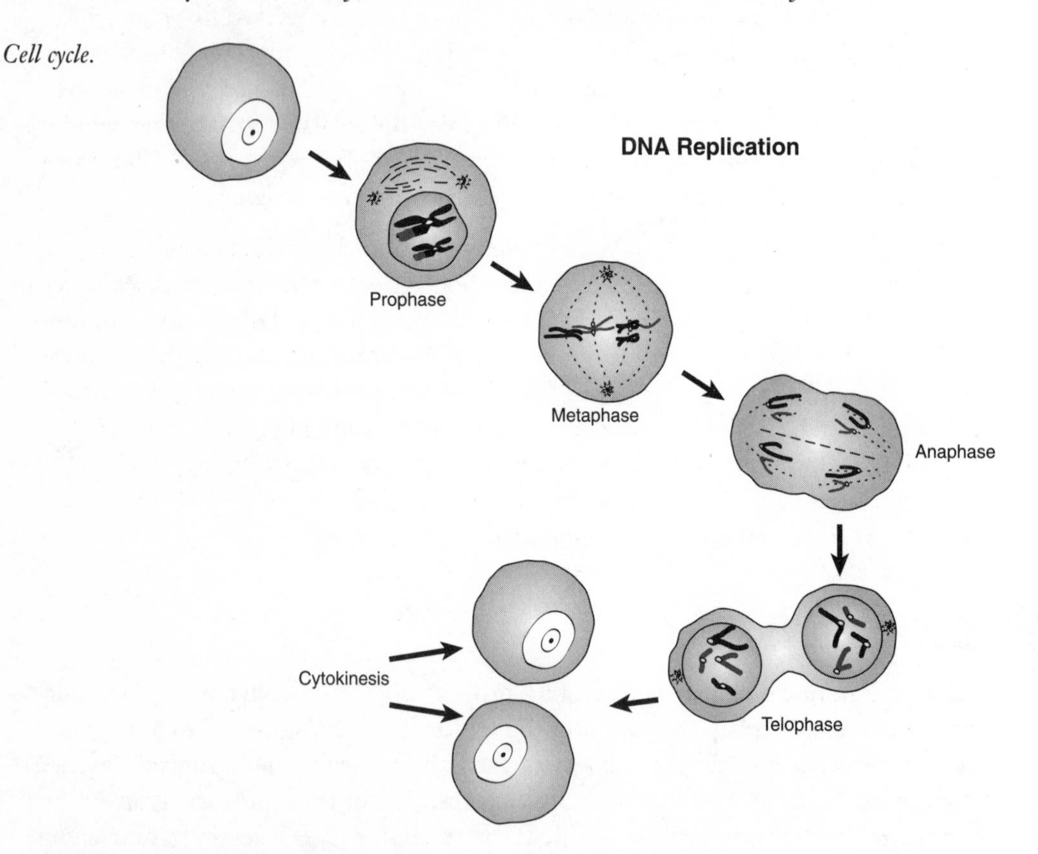

Interphase, Mitosis, Cytokinesis

The interphase continuum of stages, G_1, S, and G_2, begins the process in which the cell grows and matures (G_1), followed by the S phase in which the DNA is copied. Finally, the G_2 phase is when the cell prepares for division. *Mitosis* occurs when the nucleus of the cell divides into two identical nuclei with the same number and type of chromosomes, followed by *cytokinesis* when the cytoplasm, for both plant and animal cells, divides, thus creating two daughter cells that are genetically equal and approximately identical in size.

Cell Cycle Regulation and Cancer

Cells regulate their cell cycle in two distinct ways:

- During G_1, when the conditions are favorable, certain proteins stimulate the cell to begin copying the DNA (S phase). Likewise, if the cell is not healthy or large enough, or the environmental conditions are not favorable, the cell cycle stops here to prevent cell injury.
- The cell cycle can also cease during the G_2 phase at the DNA replication site. If the DNA is determined to be without blemish, the process continues, if damaged, the cell cycle is suspended until the DNA can be repaired.

If a mutation occurs in one of the genes that controls or regulates cell growth in any number of ways, the corresponding protein may not function correctly, allowing the cell cycle to proceed without interruption. *Cancer* is a cell-division disorder that results in uncontrolled cell growth.

CAUTION

Biohazard

Although mutations occur spontaneously in nature, environmental factors may increase their incidence, such as the use of tobacco products, overexposure to ultraviolet light and other type of radiation, and certain viruses.

Asexual Reproduction

When an organism reproduces offspring without the union of *gametes*, then asexual reproduction has occurred. Gametes are sex cells that are either sperm (male) or egg (female).

Asexual reproduction guarantees that the offspring will be both genetically and structurally identical to each other and their parent. It also allows one parent to rapidly produce offspring, which can be an ecological advantage when exploiting a new territory. For instance, the foxtail plant is considered a "weed" in certain parts of the Midwest. It is an annual plant that is able to produce enough seeds to cover an exposed area in one growing season. In one sense this is good, because it may prevent erosion; unfortunately, the area covered by the foxtail may be a farmer's field.

It also increases the likelihood that the species will survive, simply because of massive numbers. The identical nature of the offspring is also a potential drawback because of a major change in the environment, or the blanket use of a biocide, or a hungry predator.

Asexual reproduction produces offspring in four distinct methods:

- *Budding* is when offspring begin as outgrowths or "buds" of the parent. When mature, they drop off and grow into a mature adult. Budding is common in Porifera, like sponges see Chapter 20.
- *Fragmentation* is common in cnidarians and some worms, and occurs when a piece or pieces of an organism are cut off or broken off from the main body. The fragmented piece then grows into an adult.
- *Binary fission* is a combination of mitosis and cytokinesis because an organism simply divides into two organisms, especially common in flatworms.
- *Parthenogenesis* is the deposition of unfertilized eggs, often by insects, which grow into adults.

In every case, the offspring are identical to the parent and to each other.

The Least You Need to Know

- Cell size and efficiency are determined by the ratio of cell surface area to cell volume.
- Prokaryotes have no internal membranes and are simpler than eukaryotes.
- Viruses are extremely small and have two mechanisms of reproduction that create diseases for their host.
- Organisms use many different methods to move materials across membranes; some are passive because they do not require energy, others are active because ATP is required.
- The endomembrane system compartmentalizes the cell and provides sites for a variety of cellular functions.

Chapter 3

Specialized Cell Structure and Function

In This Chapter

- Structural modifications within organisms
- Chemical and energy interactions between photosynthesis and respiration
- How energy is released in a cell
- Why you inhale and exhale and what happens if you don't
- Specialized cellular events required for protein synthesis
- How DNA and RNA are involved in protein synthesis

The study of cells was divided into two chapters so that you could double your pleasure. Actually, this chapter advances some of the basic cell functions and integrates several chemical compounds. You may want to refer back to Chapter 1 as you proceed.

Both plants and animals exhibit a range of complex functions that are really several simple functions joined together. As you progress through this chapter, note how the big ones have been broken down to their basic parts or stages for easy understanding. Some of the best biological principles are included in this chapter. The secrets to photosynthesis, cellular respiration, and protein synthesis—long fortresses of unconquerable knowledge—will be subdued and laid before you.

It is not necessary to take an aspirin for an anticipated headache, or gulp another cup of coffee in preparation for a long one. This chapter is arranged for quick understanding and retention; try it and see.

Modifications and Adaptive Functions

As organisms felt the pressure of natural selection and attempted to colonize new territories, the need for advancements in both structure and function were necessary. They evolved with modified structures (evolution) or were created with the appropriate structures (creationism) to fit the environment. Advancements in structure and function created complex life-supporting systems that are more versatile and allow the organisms greater freedom for colonization in fringe territories. Photosynthesis, respiration, and protein synthesis are typical examples of complex chemical phenomena that occur around and within us constantly.

Location, Structure, Function

Most people think that *photosynthesis* is a process in which six carbon dioxide molecules react with six water molecules using the power of light to create one molecule of glucose and six molecules of oxygen. Although this simplified equation is fundamentally correct, it does not highlight the various reactions that make photosynthesis an interesting plant cellular event.

Bioterms

Photosynthesis is the process whereby plants convert light energy into chemical energy by making a variety of organic molecules, including glucose, from carbon dioxide (CO_2) and water (H_2O).

It is believed that 3 to 3.5 billion years ago, photosynthesis changed the environment of the early Earth from one that lacked oxygen to an oxygen-rich environment that, in turn, profoundly changed the type of early life that could live there and forced the extinction of many existing species.

All organisms currently living on Earth are basically solar powered because, except for a few unique creatures, all life receives energy from the food made by photosynthesis.

Photosynthesis

The mechanism for photosynthesis begins with an understanding of the location and structure of the active area in the plant. As previously described, *chloroplasts* are the organelles where photosynthesis takes place. Within the chloroplasts, the stroma is the location where carbon dioxide is converted into sugar and where the specialized membranes called *thylakoids* are located. Within the thylakoids, chlorophyll molecules actually trap and process the energy of light with the help of two photosystems described in the following section ("Light Reaction"). Carbon dioxide enters and oxygen exits the plant from adjustable openings on the underside of leaves called *stomata*. Water is drawn into the plant by its cohesive nature from roots in terrestrial plants and by direct absorption from the environment in aquatic plants. Plants also grow to maximize their exposure to light. Refer to the illustration *Chloroplasts*.

Bionote

Redox reactions are common in biological systems. A substance is oxidized when it loses electrons, whereas the substance that gains electrons is reduced. In photosynthesis, water is oxidized and carbon dioxide is reduced.

Photosynthesis is an oxidation-reduction (redox) type of reaction.

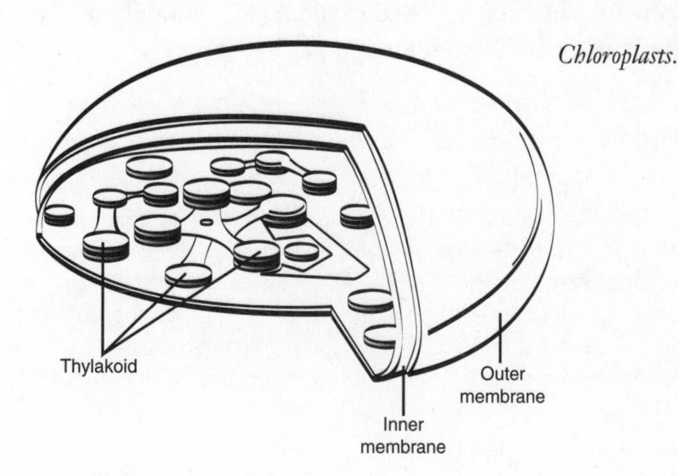

Chloroplasts.

Light Reaction

Photosynthesis is divided into two stages. The first stage, called the *light reaction*, is where the energy of light is captured by the chlorophyll molecules and processed to create high-energy compounds that are used later in the *dark reaction* (covered in the section that follows). The second stage, known as the Calvin cycle after its discoverer, is also known as the dark reaction, because it uses the energy created in the light reaction to bond carbon chains together to form sugars, other carbohydrates, proteins, lipids, and nucleic acids.

The light reaction occurs in four distinct processes that run continuously if conditions permit:

- Light energy is absorbed by chlorophyll molecules and transferred to make high-energy electrons.
- High-energy electrons enter the electron transport chain where their energy is transferred to electron acceptors.
- Water is oxidized to produce hydrogen ions and the waste gas, oxygen.
- High-energy compounds, *ATP* and *NADH*, are formed.

The mechanism for the four processes involves an interaction between structure and function. Within the thylakoid membrane are clusters of pigmented molecules (called *photosystems*), in addition to chlorophyll, that cooperate to capture and process the energy of light. There are two photosystems that contain 200 to 400 molecules of chlorophyll and other supporting pigments that collectively transfer the light energy to create a high-energy electron(s). Oddly enough, they are called photosystem 1 and photosystem 2, even though photosystem 2 usually begins the reaction.

Bionote

Supporting pigments, such as carotenoids, capture light energy of wavelengths not useful to chlorophyll molecules, as a means of extending the plant's ability to capture more of the energy from the available light. These support pigments usually operate at the lower, less-energetic wavelengths of light, and their colorful presence becomes evident in the fall after the masking chlorophyll molecules are decomposed by the shorter days and cooler weather to expose the orange, yellow, red, and purple pigments that provide the fall colors in deciduous trees (ones that drop their leaves).

When a photon of light hits the photosystem, the pigmented molecules absorb the energy and transfer it to either of two central chlorophyll molecules: P_{700}, which activates photosystem 1; or P_{680}, which activates photosystem 2. P_{700} and P_{680} reference two types of chlorophyll molecule. The *P* stands for "pigment" and the numbers refer to the wavelength of light that activates them.

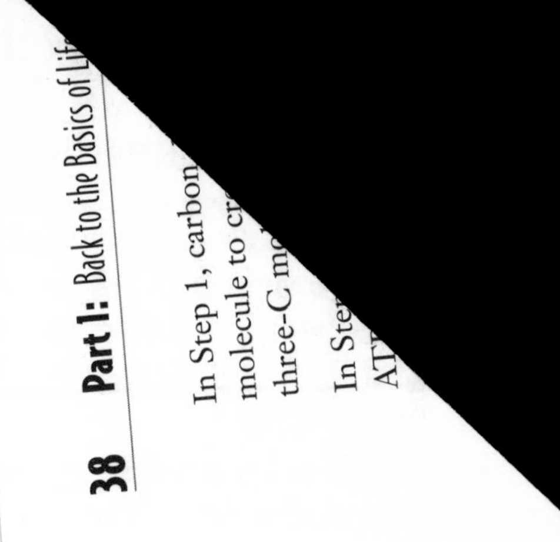

In the current model, photosystem 2 creates A
compounds. Whenever a photon of light is tra
the energy to one of its electrons. This energiz
energy level beyond the attractive force of the
to immediately be received by an electron-accep
entry into the *electron transport chain.* The P_{680} c
the electron-acceptor molecule is reduced.

The electron transport chain, located in the thyl
cules that systematically remove energy from the
to molecule. The energy subtracted from the ele
(hydrogen ions, H^+) into the thylakoid. The extr
water was oxidized) in the thylakoid membrane se potential energy gradient, much like water when impounded behind a dam. As the hydrogen ions push and return through channel protein gates in the membrane, these specialized protein gates use that kinetic energy to catalyze a phosphorylation reaction that adds a high-energy phosphate group to an ADP molecule, creating ATP.

Meanwhile, light energy is also absorbed by the P_{700} chlorophyll molecule in photosystem 1, which also transfers the energy to its electrons, which excites them to enter a different electron transport chain. The oxidized P_{700} molecule then instantly attracts the loose and energy-rich electrons created in the photosystem 2 electron transport chain to replenish its electron cloud. The excited electron from photosystem 1 combines to reduce $NADP^+$ to the energy-rich compound NADH. Refer to the illustration *Photosystem Model.*

In summary, within the light reaction, there is a continuous flow of electrons from water to photosystem 2, which creates energy-rich ATP and provides energy-depleted electrons to photosystem 1, which then replaces the excited electrons that enter a different electron transport chain to create NADH. The light reaction harnesses the energy of light and transfers it to the chemical energy of a molecule.

Calvin Cycle or Dark Reaction

The Calvin cycle is called the dark reaction because it does not need light to make biomolecules from the energy created in the light reaction. The Calvin cycle is explained in three steps:

1. Formation of PGA, a three-carbon molecule
2. Conversion of PGA into PGAL
3. Recovery of the starting material and formation of organic compounds

dioxide bonds with the five-carbon RuDP (ribulose diphosphate) ...eate a temporary six-carbon molecule that immediately splits into two, ...olecules called PGA.

...p 2, PGA receives a high-energy phosphate group from ATP (de-energizing ...P to ADP, which can then be reused in the light reaction). Next, NADH adds a proton (hydrogen ion) and releases a phosphate group, thus creating PGAL and the now energy-poor NADP molecule.

In Step 3, most of the newly created PGAL is converted into RuDP, which can then re-enter and restart the Calvin cycle. However, one out of every six PGAL molecules is converted into organic compounds needed elsewhere by the cell.

Statistically, six revolutions of the Calvin cycle are needed along with the addition of six carbon dioxide molecules to create a six-carbon sugar such as glucose. So, technically, the simplistic equation for photosynthesis is correct, however, not explanatory. A better representation is the following:

> Light reaction: Water + ADP + NADP + Phosphate + Light energy → ATP + NADH + Oxygen
>
> Calvin cycle: ATP + NADPH + RuDP + Carbon dioxide → PGAL + $NADP^+$

Alternative Carbon-Fixing Pathways

It is worth noting that two alternative mechanisms have evolved in response to different environmental conditions for converting carbon dioxide into organic compounds, called *carbon fixation:* C_4 and CAM. The C_4 method has unique enzymes that combine inorganic carbon dioxide to create a four-carbon intermediate molecule rather than the three-carbon molecules of the Calvin cycle. The C_4 mechanism is an advantage for certain types of plants because on hot, sunny days, C_4 plants partially close their stomata to minimize carbon dioxide entry and oxygen release. Normally this greatly reduces the carbon-fixation capabilities; however, C_4 plants have an additional enzyme that allows the formation of four-carbon molecules from carbon dioxide even when the carbon dioxide concentrations are low. The four-C molecules then deliver the carbon dioxide to the Calvin cycle, where it is processed. C_4 plants, such as crabgrass and corn, only have an advantage during the hot summer months because the energy cost of the C_4 mechanism is more expensive than C_3 (light reaction).

CAM mechanics are designed to conserve water for succulent plants, such as cacti and pineapples, that have adapted to live in hot, dry *biomes* (see Chapter 24). Unlike all other plants, CAM plants only open their stomata at night, when they are least likely to lose water by evaporation. Therefore, they take in and fix carbon dioxide at night and create organic compounds that release the carbon dioxide during the day to enter the Calvin cycle. Because they operate at night, their energy-trapping and absorption capacity is limited, so they grow slowly; however, they retain water very well and are therefore highly adapted to these adverse conditions. The sum total of all photosynthetic pathways is the same: the production of glucose (and other organic compounds).

Bionote

Aerobic cellular respiration is the exact opposite of photosynthesis. The starting products of either reaction are the end product of the other, and vice versa. They are balanced, opposing reactions.

Cellular Respiration

Cellular respiration harvests the energy created during photosynthesis through a series of biochemical reactions:

- Glycolysis
- Kreb's cycle
- Electron transport chain
- Fermentation

Glycolysis

Cells harvest the energy contained in the chemical bonds of glucose in a very controlled, step-by-step series of reactions that release small amounts of energy during each biochemical reaction. *Glycolysis* is the first step of cellular respiration, where a molecule of glucose is split to release energy.

Bionote

Historically it is thought that glycolysis was the first method of energy generation by early-Earth organisms because of the availability of glucose and the lack of an oxygen environment.

The process of glycolysis is an enzyme controlled, four-step reaction that occurs in the cytoplasm of the cells:

1. Energy is required to begin the process, so a molecule of glucose accepts two high-energy phosphate groups from two ATP molecules.

2. The resulting intermediary molecule immediately divides into two, three-carbon molecules called PGAL, each containing a high-energy phosphate group.
3. A second high-energy phosphate group is added to the three-carbon PGAL molecule and two NADH molecules are produced.
4. Finally, the three-carbon PGAL molecules donate their high-energy phosphate to create ATP and the three-carbon pyruvate forms as the final products.

Refer to the illustration *Glycolysis* to view the action of the intermediary compounds and the energy molecules.

Glycolysis.

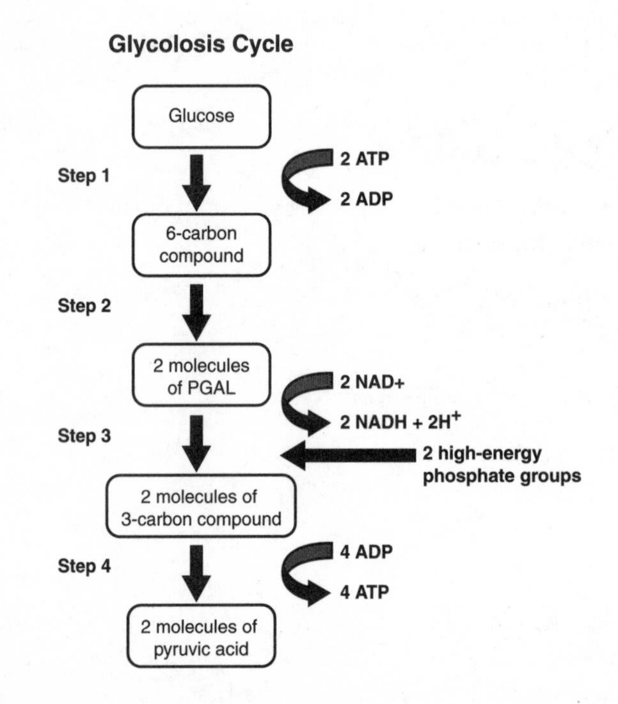

The high-energy phosphate groups added in Steps 1 and 3 are removed in Step 4 to create four ATP (two from Step 1 and two from Step 3) from four ADP. Because two ATP molecules were required to begin glycolysis and four were produced in the final step, the net gain is two ATP molecules. Although ATP is produced, glycolysis does not produce enough energy to sustain their life cycles for complex life forms. Therefore, the main purpose of glycolysis is to produce high-energy electrons for use in the electron transport chain. The final product, pyruvate or pyruvic acid, still contains energy that can be harvested in two ways depending on the availability of oxygen.

Kreb's Cycle

If oxygen is present, pyruvate undergoes *aerobic respiration*, which consists of two parts: the Kreb's cycle (also known as the citric acid cycle), and the electron transport chain. Following glycolysis, pyruvate moves from the cytoplasm into the mitochondria and reacts with a coenzyme to create the two-carbon molecule, acetyl coenzyme A (acetyl CoA) by losing one carbon dioxide molecule. Nine enzyme-controlled reactions are condensed and presented in the following five steps:

1. Acetyl CoA donates the two-carbon acetyl group to a four-carbon intermediary compound, oxaloacetic acid, to create the six-carbon citric acid molecule.
2. The high-energy electrons are oxidized to create the energy-rich NADH molecule when the six-carbon compound loses a carbon dioxide molecule to become a five-carbon molecule.
3. A second molecule of NADH and a molecule of ATP are produced when another carbon dioxide molecule is released from the five-carbon molecule, which then degrades to a new four-carbon molecule.
4. The four-carbon molecule is further oxidized to transfer high-energy electrons to create the high-energy compound, $FADH_2$, and more NADH.
5. Enzymes rearrange bonding within the four-carbon molecule to become oxaloacetic acid, which combines with the acetyl CoA to restart the Kreb's cycle.

Refer to the illustration *Kreb's cycle* to see the breakdown of the energy-containing molecules.

In summary, the Kreb's cycle removes carbon dioxide molecules from glucose in a stepwise fashion to release energy, but like glycolysis, the primary purpose is to create the high-energy electron carriers NADH and $FADH_2$. The carbon dioxide expelled in the process is a waste product and must be removed from the system. For example, this is why you exhale.

The Kreb's cycle produces 10 ATP molecules and generates the energy molecules NADH and $FADH_2$, which are harvested later in the *electron transport chain.*

At the conclusion of the Kreb's cycle, the original glucose molecule is completely oxidized, so most of the energy now resides in the high-energy electrons removed from the carbon-carbon and carbon-hydrogen bonds that created the electron carriers NADH and $FADH_2$. The high-energy electrons contained in NADH and $FADH_2$ are donated to the electron acceptor molecules located on the long folds of *cristae* on the inner membrane of the mitochondria. They begin the process of ATP formation in the electron transport chain.

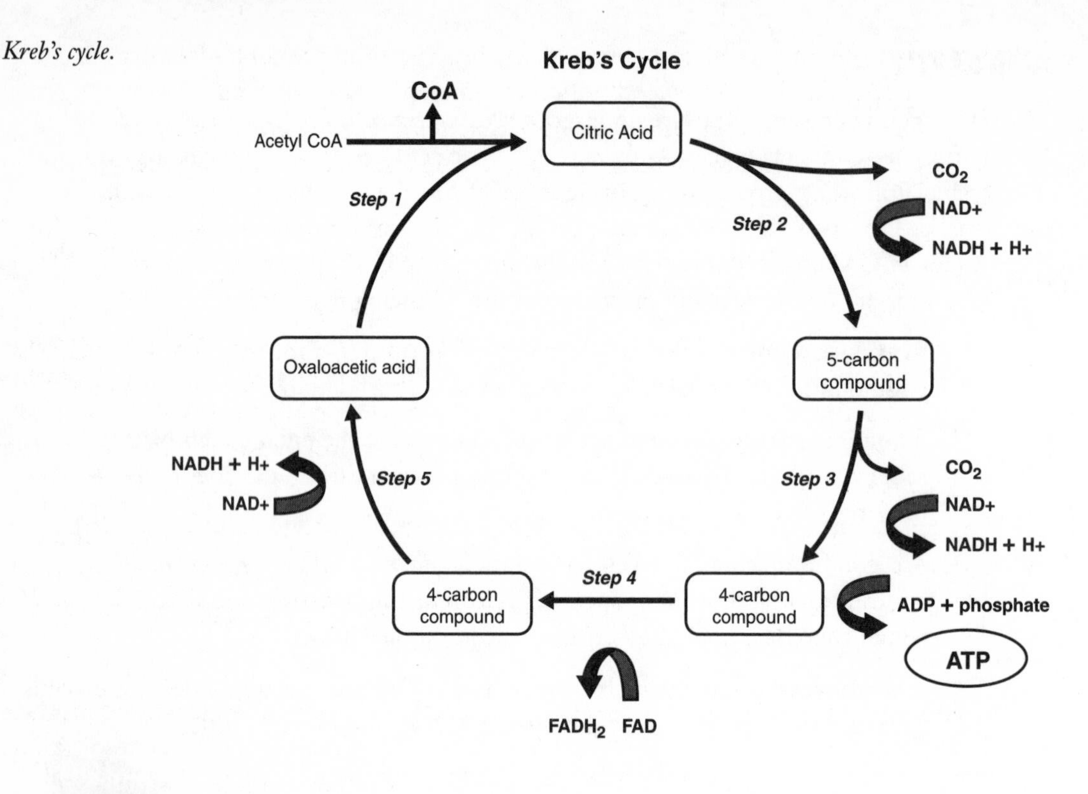

Kreb's cycle.

Bioterms

The **electron transport chain** is a series of molecules called *cytochromes* and associated enzymes that pass high-energy electrons from molecule to molecule, removing energy in a stepwise mechanism. The last acceptor of the now energy-depleted electron is oxygen, which then combines with the excess hydrogen ions from the cytoplasm to create water. This is why you inhale.

Electron Transport Chain

The electron transport chain is where most of the energy is released in cellular respiration. The mechanism of the electron transport chain can be described in five steps:

1. High-energy electrons from NADH and $FADH_2$ enter the electron transport chain and are passed from molecule to molecule, losing energy in a controlled stepwise manner.
2. The energy lost from the electrons is used to pump hydrogen ions from the inner mitochondrial compartment to the outer mitochondrial compartment

across the mitochondrial membrane. This creates an area of high hydrogen ion concentration on one side of the mitochondrial membrane and a low hydrogen ion concentration on the other side of the membrane. The result is a concentration gradient across the inner membrane creating a source of potential energy, which is again comparable to the potential energy of water held back by a giant dam.

3. The concentration gradient is used as a source of potential energy to drive the *chemiosmotic* synthesis of ATP.
4. A carrier protein helps the hydrogen ions diffuse through a channel protein opening in the membrane. As the hydrogen ions diffuse from the area of high concentration to the area of low concentration, the carrier protein harnesses the kinetic energy of the hydrogen ion to add a high-energy phosphate group to ADP forming ATP, with the help of the enzyme ATP synthetase.
5. The high-energy electron is passed along the electron transport chain until the excess energy is removed and then it is combined with the excess hydrogen ions and oxygen to form water, which then becomes a waste and must be removed from the system. For example, this is why you urinate.

In summary, the oxidation of glucose is approximately 37 percent efficient and produces all the energy required for almost every type of cell. The complete aerobic respiration of 1 molecule of glucose creates a maximum yield of 36 ATP molecules, as follows:

- Glycolysis = 4 ATP molecules
- Kreb's cycle = 10 ATP
- Electron transport chain = 22 ATP

Fermentation

If oxygen is not present after glycolysis, the electron transport chain cannot operate because there is no oxygen present to serve as the final electron acceptor. So the pyruvate is converted by certain specialized cells into other compounds in a process called *fermentation*. Fermentation does not produce additional ATP, but it does regenerate NAD^+, which can then participate in glycolysis to make more ATP. The NADH is converted to NAD^+ by adding the extra high-energy electron in NADH to an intermediate organic molecule. The combined total of glycolysis and fermentation produces 2 ATP molecules for every glucose, compared with 36 ATP via aerobic respiration. Although there are several fermentation pathways, the two most common produce lactic acid and ethanol.

In the enzyme-catalyzed *lactic acid* fermentation, the three-carbon pyruvate is rearranged into the three-carbon molecule lactate, a.k.a. lactic acid. In the process, NADH is oxidized to NAD^+, which is then available for use in glycolysis, while pyruvate is reduced to lactate. This process is familiar to athletes because excess lactic acid buildup due to anaerobic exercise (not receiving enough oxygen), causes painful areas in the affected muscles.

Alcohol fermentation produces *ethanol*, also known as ethyl alcohol, the alcoholic component of adult alcoholic beverages. This two-stage process begins when pyruvate loses a carbon dioxide molecule to become an intermediary two-carbon molecule. In Step 2, the hydrogen ions from NADH are added to create ethyl alcohol and regenerate the NAD^+.

Protein Synthesis

The making of the various types of protein is one of the most important events for a cell because protein not only forms structural components of the cell, it also composes the enzymes that catalyze the production of the remaining organic biomolecules necessary for life. In general, the genotype coded for in the DNA is expressed as a phenotype by the protein and other enzyme-catalyzed products.

The DNA housed in the nucleus is too large to move through the nuclear membrane, so it must be copied by the smaller, single-stranded RNA (transcription), which moves out of the nucleus to ribosomes located in the cytoplasm and rough endoplasmic reticulum to direct the assembly of protein (translation). The genes do not actually make the protein, but they provide the blueprint in the form of RNA, which directs the protein synthesis.

Transcription

Transcription occurs in the cell nucleus and represents the transfer of the genetic code from DNA to a complementary RNA. The enzyme RNA polymerase …

- Attaches to and unzips the DNA molecule to become two separate strands.
- Binds to *promoter* segments of DNA that indicate the beginning of the single strand of DNA to be copied.
- Moves along the DNA and matches the DNA nucleotides with a complementary RNA nucleotide to create a new RNA molecule that is patterned after the DNA.

The copying of the DNA continues until the RNA polymerase reaches a *termination signal*, which is a specific set of nucleotides that mark the end of the gene to be copied and also signals the disconnecting of the DNA with the newly minted RNA.

The three types of RNA are …

- mRNA (messenger RNA) is transcribed from DNA and carries the genetic information from the DNA to be translated into amino acids.
- tRNA (transfer RNA) "interprets" the three-letter codons of the nucleic acids to the one-letter amino acid word
- rRNA (ribosomal RNA) is the most abundant type of RNA, and along with associated proteins compose the ribosomes.

When the RNA polymerase is finished copying a particular segment of DNA, the DNA reconfigures into the original double-helix structure. The newly created mRNA moves out of the nucleus and into the cytoplasm.

Translation

Translation is the conversion of information contained in a sequence of mRNA nucleotides into a sequence of amino acids that bond together to create a protein. The mRNA moves to the *ribosomes* and is "read" by tRNA, which analyzes sections of three adjoining nucleotide sequences, called *codons*, on the mRNA and brings the corresponding amino acid for assembly into the growing polypeptide chain. The three nucleotides in a codon are specific for a particular amino acid. Therefore, each codon signals for the inclusion of a specific amino acid, which combines in the correct sequence to create the specific protein that the DNA coded for.

The assembly of the polypeptide begins when a ribosome attaches to a *start codon* located on the mRNA. Then tRNA carries the amino acid to the ribosomes, which are made of rRNA and protein and have three bonding sites to promote the synthesis. The first site orients the mRNA so the codons are accessible to the tRNA, which occupy the remaining two sites as they deposit their amino acids and then release from the mRNA to search for more amino acids. Translation continues until the ribosome recognizes a codon that signals the end of the amino acid sequence. The polypeptide, when completed, is in its primary structure. It is then released from the ribosome to begin contortions to configure into the final form to begin its function.

After the proteins are made, they are packaged and transported to their final destination in an interesting pathway that can be described in three steps involving three organelles:

Bionote

Each codon on the mRNA specifies a particular amino acid, which is recognized by the anticodon of the complementary tRNA. There are 20 different amino acids; there are also 20 different tRNA molecules.

1. *Vesicles* transport the proteins from the ribosomes to the *Golgi apparatus*, a.k.a Golgi complex, where they are packaged into new vesicles.
2. The vesicles migrate to the membrane and release their protein to the outside of the cell.
3. *Lysosomes* digest and recycle the waste materials for reuse by the cell.

Enzymes within the Golgi apparatus modify the proteins and enclose them in a new vesicle that buds from the surface of the Golgi apparatus. The Golgi apparatus is often seen as the packaging and distribution center of the cell.

Vesicles are small, membrane-enclosed envelopes that are usually made in the endoplasmic reticulum or Golgi apparatus and are used to transport substances through the cell.

Lysosomes are a special type of vesicle that contains the digestive enzymes for the cell and are useful in breaking down leftover waste products of proteins, lipids, carbohydrates, and nucleic acids into their component parts for reassembly and reuse by the cell.

The Least You Need to Know

- The light reaction of photosynthesis captures the energy of light; the Calvin cycle uses that energy to create organic molecules, such as glucose.
- Most of the energy captured as a result of photosynthesis is released by high-energy electrons during the electron transport chain with smaller amounts released in the Kreb's cycle and glycolysis.
- Alternative energy-capturing mechanisms to photosynthesis have evolved to allow certain organisms to colonize harsh environments.
- Photosynthesis and cellular respiration are complementary reactions.
- Most cellular respiration pathways include oxygen, but several pathways are available to organisms that allow them to live in oxygen-poor or -deficient environments.
- The extraordinary process of making the correct protein involves reading, coding, and transcribing DNA to correctly assemble the appropriate amino acid.

Part 2

Genetics and Inheritance

This section includes some of the latest activity in genetics, genetic engineering, and common inheritance patterns. Your newfound knowledge will make you the conversational highlight at your next party. Everyone will be amazed at how much you know about the Human Genome Project, current recombinant cloning technologies, and the problems associated with them. Inheritance models are provided with explanations of how they occur, including some that may fit your pedigree, family tree, or phylogeny, you see?

Inheritance

In This Chapter

- Mendel's ideas on heredity
- Why genetic abnormalities can be a good thing
- How genes turn on and off
- How sex cells are produced in plants and animals

Most people like this topic of inheritance. It references some things that are obvious in our everyday life, but also includes items that we didn't know much about until recently.

Most of this chapter talks about Mendel's work with garden peas and how much information people can learn from their garden if they take careful notes and a solemn oath to tend it for many years in a row.

Then we move on to an interesting conundrum: the production of sperm and egg is similar but different; gene regulation in prokaryotes and eukaryotes is similar but different; but the mechanism of meiosis is the same for everybody. Read on to find out how this applies to you and your everyday life.

And last, but not least … mutations, which can be very interesting. Perhaps you have some. They are the only real source of genetic variation and explain how certain species can survive in a changing environment. Alternation of generations may seem like a plant mutation, but it isn't. In fact, it helps certain plants survive. We don't alternate our generations; we plow right on through.

Genes at Work: Mendel's Breakthrough

Hippocrates hypothesized that particles called *pangenes* occur in all parts of the body, and change as a person matures to reflect body changes, and then enter the egg or sperm of an individual to be passed on to the next generation. Although this idea seemed to make sense and was widely held as correct at the time, it unfortunately did not account for genetic abnormalities or even common human discrepancies such as hair- or eye-color variations.

Pangenesis was replaced in the early 1800s by the hypothesis of "blending," which stated that offspring inherit and blend traits from their parents. For instance, a red flower crossed with a white flower should always produce a pink flower. Unfortunately, this is not always the case; sometimes a red flower and a white flower produce a red flower or a white flower, not necessarily pink.

A student at the University of Vienna published a paper in 1866 that became the foundation for modern genetics. Gregor Mendel, an Augustinian monk, grew and investigated garden peas and subsequently explained "heritable" factors. He also posed and confirmed several salient hypotheses and instituted a new vocabulary to explain his findings that is still in use today.

There is more than one form for a gene (allele). For example, based on his experimentation with purple and white flowering pea plants, he deduced after examining the results of numerous crossings, that one gene is responsible for the purple color and a different gene is responsible for the white color of the flowers.

In most cases, two genes are needed to express a trait, one from each parent. Likewise, if an organism is *true-breeding*, both genes that control the same trait had to be the same gene (*homozygous*).

The principle of segregation states that during egg and sperm formation, the pair of genes separates (segregates) and only one gene becomes incorporated into either the sperm or the egg, and the other gene can also become incorporated into either a sperm or egg, so that when sperm and egg unite, the offspring has the normal number, or pair, of genes.

When two different alleles for the same trait are found in an organism, the gene that is expressed is the *dominant allele*, and the gene not expressed is the *recessive allele*. If red is dominant over white in flowers, the offspring between a homozygous red and white flower should all be red.

The principle of independent assortment states that each gene from the pair of alleles separates independently during gamete formation. Mendel noticed that during a *dihybrid cross* (cross of two traits), the offspring inherited the alleles as if they separated from each other and had equal chance of inheritance as the alleles from the other trait. Specifically, Mendel crossed a plant that was homozygous dominant for round, yellow seeds (RRYY, R = round, Y = yellow) with a homozygous plant with wrinkled, green seeds (rryy, r = wrinkled, y = green). If the genes stuck together and did not act independently, he hypothesized that the *F_2 generation* must contain the same genotype as the parents, either RY or ry. For reference purposes, the F_1 generation is the product of the mating of the original parents; the F_2 is the product of the F_2 mating. The F_2 then becomes the "grandchildren" of the original mating. Through experimentation he found the following genotypes: RY, Ry, rY, ry. He correctly concluded that the individual alleles were segregating from the paired alleles and were free to unite in all possible combinations. Refer to the illustration *F_1 Punnet square*. It can be used to predict the results of an RY and ry mating.

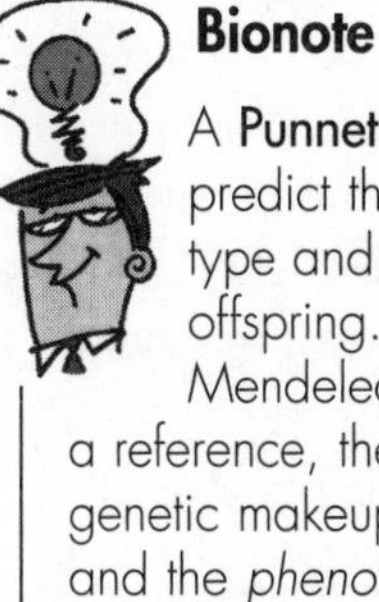

Bionote

A **Punnet square** is used to predict the possible genotype and phenotype of offspring. It operates on Mendelean principles. As a reference, the genotype is the genetic makeup of an individual and the *phenotype* is the expression of the genetic makeup. In other words, the phenotype is what the organism looks like.

First, the genotype of the parents must be known. In this case they are RRYY and rryy. Next they are segregated. If they independently sort, the possible genotypes are RY and ry.

	ry
RY	RrYy

F_1 Punnet square.

All of the F_1 (first generation) offspring are heterozygous, meaning they contained both a dominant and recessive gene, RrYy, round and yellow. However, in the F_2 generation, interesting data revealed a variety of offspring. Refer to the illustration *F_2 Punnet Square*. Parent's genotype: RrYy × RrYy.

	Ry	**ry**	**RY**	**rY**
Ry	RRyy	Rryy	RRYy	RrYy
ry	Rryy	rryy	RrYy	rrYy
RY	RRYy	RrYy	RRYy	RrYY
rY	RrYy	RrYy	RrYY	rryy

F_2 Punnet square.

Offspring phenotype:

- Round yellow = 9, round green = 3
- Wrinkled yellow = 3, wrinkled green = 1

The gametes had to sort independent of each other as depicted. The phenotypic ratio of 9:3:3:1 is typical of a dihybrid cross.

Mendel could tell by the phenotypes of the offspring that the gametes had recombined in such a way that they had to have an equal chance of combining with the other allele.

A *test cross* determines an unknown genotype by mating it with a homozygous recessive. This works for two reasons: First, the homozygous organism is an obvious true breeder. Second, the proof is in the offspring; if a homozygous recessive crosses with a homozygous dominant, all offspring are dominant; if matched with a heterozygous mate, some of the offspring will exhibit dominant and other recessive traits; a cross with another homozygous recessive produces only homozygous recessive offspring.

Test Cross

The following are the three possible test crosses that can occur. The researcher simply has to examine the phenotype to determine the unknown genotype of one of the parents. Let's simplify and only consider one gene: G is dominant for green; g is recessive for white. In every test cross, the unknown parent is always mated with the homozygous recessive parent. In this case, the test-cross parent's genotype is gg. There are three possible genotypes for the unknown parent: GG, Gg, and gg. Consider the following matings:

- In Case 1, all of the offspring are green. So when a test cross gives offspring that exhibit the dominant phenotype, the unknown parent must have been homozygous dominant. Refer to the illustration *Case 1 (all green).*

Case 1 (all green).

CASE 1

	g	g
G	Gg	Gg
G	Gg	Gg

- In Case 2, half of the offspring exhibit the dominant phenotype, and the other half exhibits the recessive phenotype. When the offspring of a test cross exhibit a 50:50 ratio of dominant to recessive, the parent had to be heterozygous. Refer to the illustration *Case 2 (half green, half white).*

CASE 2

	g	g
G	Gg	Gg
g	gg	gg

Case 2 (half green, half white).

- In Case 3, all offspring are white. The only way this could happen is if the unknown parent was homozygous recessive. Refer to the illustration *Case 3 (all white).*

CASE 3

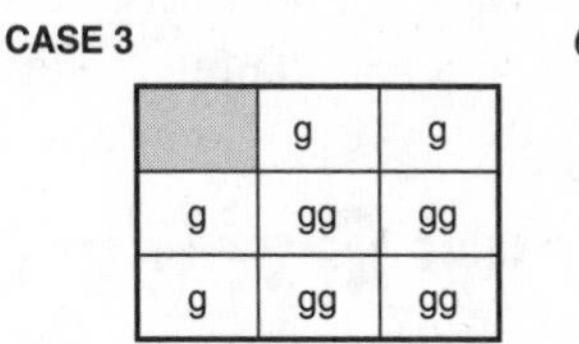

	g	g
g	gg	gg
g	gg	gg

Case 3 (all white).

A test cross is an easy way to experimentally determine the unknown genotype of a parent. But remember, the genotype of the other parent must be homozygous recessive.

Pedigree

Pedigrees are useful in tracking traits through ancestors and for predicting traits for descendents. Refer to the illustration *Pedigree.*

Assume that a "**" behind the name indicates homozygous dominant for a fictitious trait. "*–" is heterozygous, and "--" is homozygous recessive for that trait.

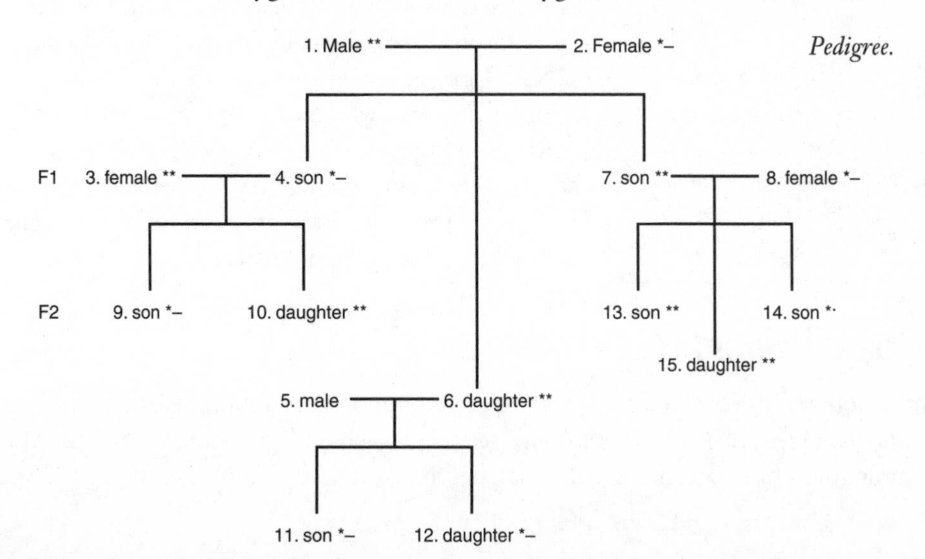

Pedigree.

The male and female mate and produce two sons, numbers 4 and 7, and a daughter, number 6 in the F1 generation. They each mate and produce the offspring shown in F2. An analysis of this pedigree shows that the fictitious trait is inherited by normal genetic inheritance patterns. For instance, the gene is not linked to the X chromosome (see Chapter 6).

Pedigrees for pets are common to ensure that the new puppy will grow into the type of dog the owner intended to purchase. Human pedigrees are also common, but in a different way. Genetic counseling provides potential parents with the likelihood that their children will inherit a specific disease. Also, most medical doctors ask as a routine question whether a history of heart disease or cancer exists in your family tree. Both are examples of the usefulness of a family pedigree.

Meiosis and Sexual Reproduction

Sexual reproduction is the union of male and female gametes to form a fertilized egg, or *zygote*. The resulting offspring inherit one half of their traits from each parent. Consequently they are not genetically identical to either parent or siblings, except in the case of identical twins. As hypothesized by Mendel, adults are *diploid*, signified as 2N, having two alleles available to code for one trait. The gametes must be *haploid*, signified by N, containing only one allele; so that when two haploid gametes combine, they produce a normal diploid individual. The process where haploid sex cells are created from diploid parents is called *meiosis*, and it occurs only in the reproductive organs.

A diploid cell undergoing meiosis first duplicates itself and then divides two times, creating four haploid cells. Meiosis begins with the same G_1, S, and G_2 stages as mitosis and also ends with a duplicate set of chromosomes. In both processes, the cell divides to form two diploid (2N) offspring; however, meiosis continues with another division, which creates the four haploid gametes, in a process called *gametogenesis*. In meiosis, several interesting events may happen along the way to provide *genetic recombination*, an unexpected change in the hereditary genetic material.

Bioterms

Gametogenesis occurs only in the ovaries and testes and represents the formation of haploid egg and sperm as a result of meiosis.

Mechanism of Meiosis I

The functional difference between mitosis and meiosis occurs in meiosis I. A synapsis occurs during prophase I, where *homologous chromosomes* align next to each other. Homologous chromosomes are the matched pair found in a diploid cell. The maternal and paternal homologous chromosomes are made of two sister *chromatids* that are duplicated copies; so at synapsis, four chromatids are aligned together in a structure called a *tetrad*, which is a fundamental difference between mitosis and meiosis. In metaphase I, the chromosome tetrads are oriented along the metaphase plate, the equator between the two opposite ends of the cell. In anaphase I, the tetrads split, with the sister chromatids remaining joined at their *centromere*. When the chromosomes arrive at their respected sides of the cell, in telophase, cytokinesis begins and one cell becomes two. They remain duplicates, but they are still haploid. The two cells now enter meiosis II. Refer to the illustration *Meiosis I* for a pictorial representation.

Bionote

A chromosome contains numerous genes and is made of two sister **chromatids** joined by a centromere. The **centromere** is the region of a chromosome where the two sister chromatids are joined. **Homologous chromosomes** are inherited from each parent and are the two chromosomes that make up a pair in a diploid cell. They are normally the same length, contain similar genes in the same location, and have a centromere at the same locus.

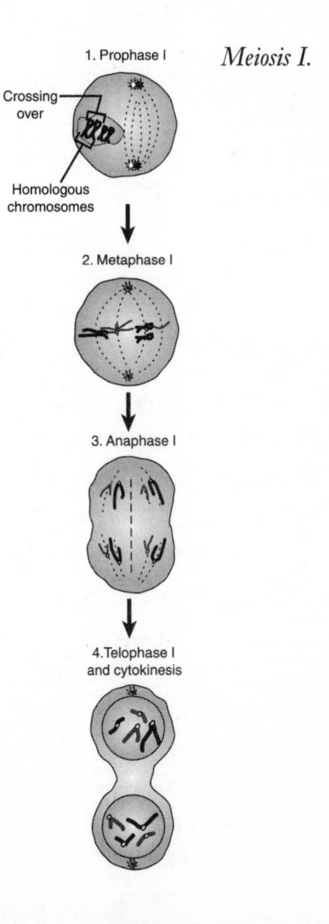

Meiosis I.

Mechanism of Meiosis II

The overall result of meiosis II is to create four haploid sex cells from the two diploid cells that began the process; refer to the illustration *Meiosis II* for a pictorial representation. Like mitosis, in metaphase II the chromosomes line up along the cell equator, and the paired chromatids separate in anaphase II. This final separation reduces the chromosome number by one-half, creating the haploid sperm and egg. Because of segregational and independent assortment, they may contain completely different alleles.

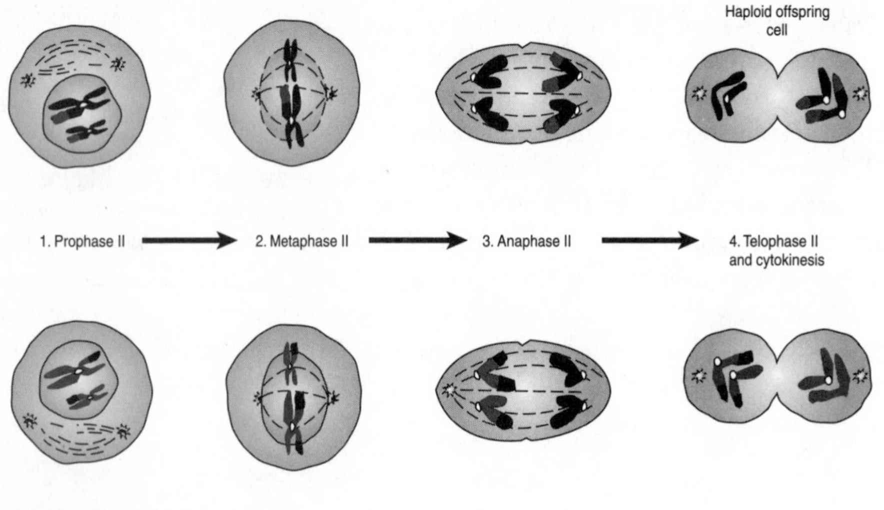

Meiosis II.

Abnormalities, Genetic Recombination, Variability

Spontaneous mistakes occur during meiosis that lead to gametes with unusual changes in their genetic structure (makeup). These gene changes lead to an unexpected *genetic recombination* that, if the organism survives, increases the genetic variability for the population. There are opportunities described next to increase genetic variability during meiosis.

The *crossing over* of sister chromatids sometimes occurs when they are aligned as tetrads in metaphase I. One chromatid or chromatid piece mistakenly lies on top of a neighboring chromatid. The neighboring nonsister chromatid absorbs the new piece of chromatid into the chromosome and releases the corresponding piece to be absorbed by the first chromosome. The net result is a new genetic recombination, because neighboring chromosomes have exchanged pieces of chromatid that will undergo meiosis as a new component of the chromosome. Refer to the illustration *Crossing over.*

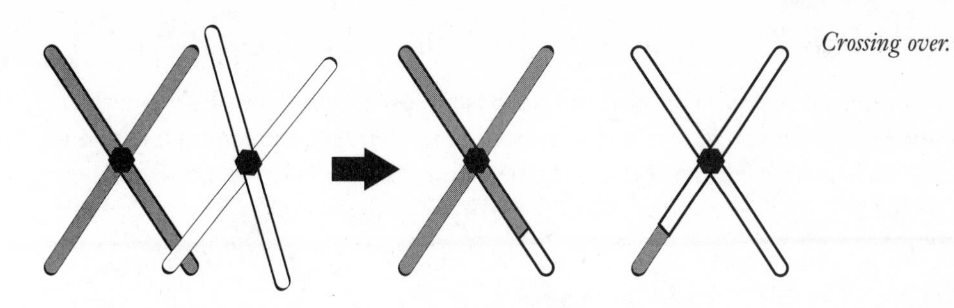

Crossing over.

The random alignment of chromosomes during metaphase I allows equal opportunity for a particular chromosome to migrate into a cell. This type of *independent assortment* gives rise to exponential gene combinations in the offspring.

Sometimes the spindle fibers fail to separate homologous chromosomes during anaphase I, which overloads one cell with chromosomes and short stocks the other. Likewise, in anaphase II of meiosis II, if a pair of sister chromatids fails to separate and migrates into the same cell, that cell now has too many chromosomes and the other, too few. These scenarios are examples of *nondisjunction*, which results in the production of gametes with an odd number of chromosomes. Because it often occurs in meiosis, the genetic recombination only affects the X and Y chromosomes, the chromosomes most noted for determining the sex of the offspring, giving rise to the following abnormalities:

- XXy = the offspring is a male with Kleinfelter's syndrome (also includes XXXy, XXXXy, and XXyy; the appearance of a single y chromosome apparently is enough to create a male). Individuals with Kleinfelter's syndrome usually display lanky builds with feminine characteristics such as breast development and poor facial and chest hair growth, and they are mentally retarded and sterile.
- Xyy = the offspring is a normal male, often called a supermale because the presence of an extra y chromosome may contribute to characteristics of increased height, weight, muscular bulk, and aggressiveness
- XXX = The offspring are female, called metafemales or superfemales, and appear normal.
- XO = The offspring are female and have Turner's syndrome (only one chromosome present). Individuals with Turner's syndrome are sterile females that are short in stature, do not sexually mature, and have a thickened web of skin between the shoulders and neck.

Mutations are the primary source of genetic variability because a mutation creates a new gene. Variability is also increased in other ways, such as by the randomness of the union between sperm and egg leading to fertilization.

Mutation

A mutation is a unique type of abnormality and is the greatest source of genetic variability because it creates a change in the nucleotide sequence composing the DNA. Mutations can be either good or bad.

Assuming the daughter cells receive the correct number of chromosomes, problems may arise in the structure of the DNA itself. Mutations involving the rearrangement of the DNA nucleotides is caused in four distinct ways.

A *translocation* occurs when the DNA double helix is broken and a piece of the chromosome attaches to a neighboring nonhomologous chromosome, making it longer than its homologous chromosome. The donating chromosome is obviously now shorter that its homolog. Refer to the illustration *Translocation.*

Translocation.

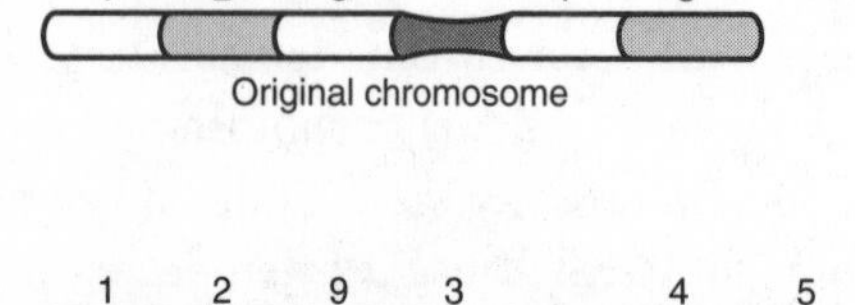

Whenever a segment of a chromosome is broken off and lost, the resulting *deletion* has serious effects on the transmission of the original genetic material. Refer to the illustration *Deletion.*

Deletion.

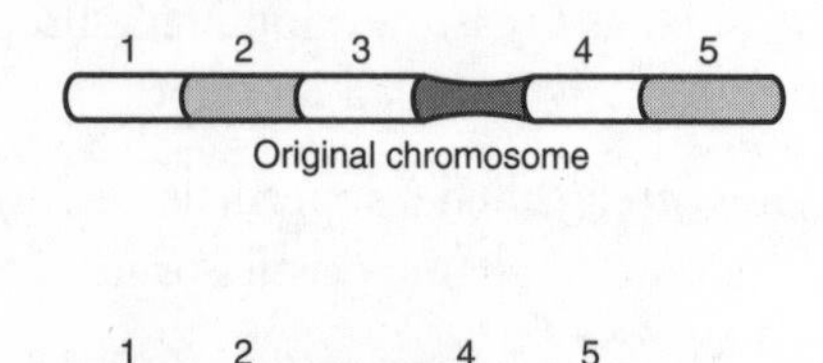

If a deleted segment returns and joins with a homologous chromosome, a *duplication* of genes has occurred. Refer to the illustration *Duplication.*

Duplication.

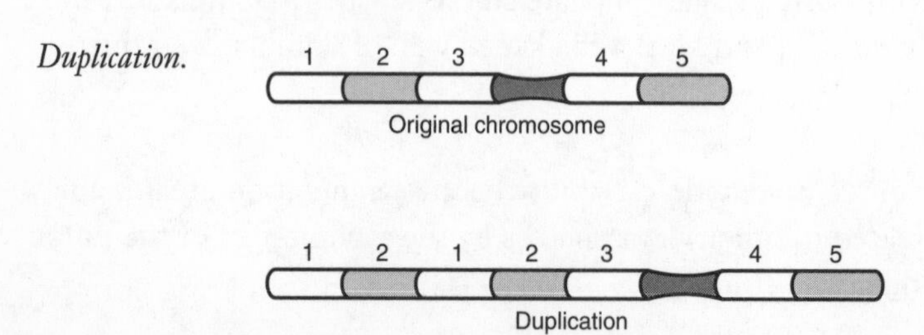

Finally, if a segment breaks loose, reverses, and reattaches in reverse order, an *inversion* results. Refer to illustration *Inversion*.

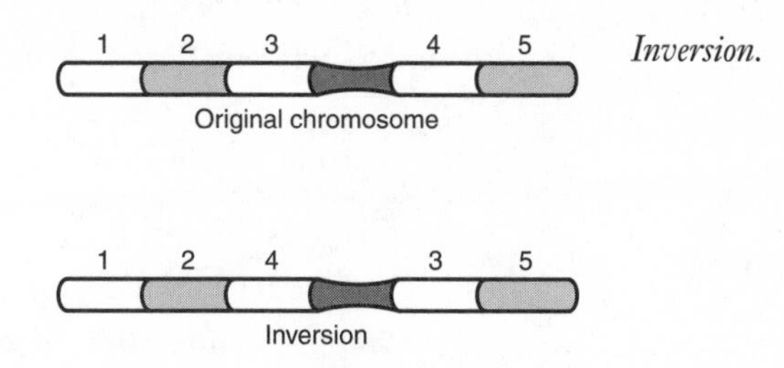

Inversion.

As a result of inversions, the genes are still present and the gene number is still the same; in translocations and deletions, however, the resulting mutations may create serious problems, especially in the case of deletions because the "reading" of the genetic code will be altered by an extra omitted gene.

On a smaller scale, a point mutation in a gene is the single exchange of one nucleotide for another. This type of genetic recombination may be too small to affect the overall function of the protein and may not be noticed by the individual, especially if it is an *interon*, which are described in a Chapter 4. In other cases, a point mutation may improve the organism by making it more fit, or make it worse by decreasing the fitness thereby lowering its chance of survival.

Bionote

A *karyotype* is a display of an individual's chromosomes that have been stained for easier observation. In humans, the karyotype shows any alteration in the 22 *autosomal* (genes that code for the body) chromosome pairs or the one pair of sex chromosomes.

Gametogenesis in Plants and Animals

Mendel hypothesized that an individual inherits an allele from each parent for each trait. In animals, gametogenesis occurs in the reproductive organs via meiosis. In humans, each sex cell is haploid, which combine to form a diploid zygote with 23 pairs or 46 total chromosomes, of which 44 are autosomes that code for traits not sex-related. The remaining pair of chromosomes, the X and y, determines the sex of the individual and control certain traits found on the individual, called *sex-linked traits* (see Chapter 6).

Oogenesis and Spermatogenesis

Oogenesis, the production of eggs in animals, involves a unique process. The diploid reproductive cell undergoes meiosis to produce one egg cell or ovum. During both cytokinesis I and II, the division of the egg cell is uneven, creating one large egg cell and three cytoplasm-deprived cells, called *polar bodies*, which are degraded and recycled by the cell. The remaining large cell develops into a mature egg cell. Conversely, in *spermatogenesis*, the diploid reproductive cell divides via the normal meiotic route to create haploid *spermatids*, all of which mature into normal sperm cells.

Bioterms

Gametangia in plants are structures that produce and create an acceptable, usually moist, environment for the site of fertilization and embryo development.

Curiously, most fungi reproduce both asexually and sexually, which is a typical sign of a species using the *alternation of generations* life cycle. In this life cycle, the sexual pathway involves a plus (+) and a minus (–) mating type meeting and joining gametangia where the haploid gametes combine to form diploid zygotes. Meiosis then occurs in the zygote, reducing it to a haploid stage for the remainder of the life cycle.

Bioterms

A **sporophyte** results from the fusing of two haploid gametes to become a diploid, spore producer; a gametophyte is a haploid, gamete producer that grows from a spore. A pine tree is an example of the most recognizable sporophyte form.

Alternation of Generations

Except for our reproductive organs, our bodies are diploid; however, plants vary greatly from animals because their life cycles include both a haploid and a diploid cycle for prolonged periods of time. Accordingly, male and female haploid *gametophytes* produce gametes by mitosis; diploid *sporophytes produce spores* by meiosis.

Gametophytes, Then Sporophytes

Gametophyte haploid gametes combine in the gametangia to produce a fertilized diploid egg. The diploid offspring divides mitotically and grows into maturity as a sporophyte and then undergoes meiosis to produce haploid spores that grow into haploid gametophytes.

In modern times, the overwhelming majority of plant species, including all seed plants, exhibit a dominant and usually recognizable sporophyte stage in their life cycle. An exception is moss. The green velvety growth that is often seen in dense shade is actually the gametophyte stage of the life cycle. When the male and female gametophyte gametes join in the gametangia, the resulting zygote develops into the sporophyte stage, which remains attached to the gametophyte by growing a stalk with a capsule full of spores on top.

Regulation of Gene Expression in Prokaryotes and Eukaryotes

Clever mechanisms turn genes off and on so that they only function when there is a need for their services.

Prokaryotes and the Operon Model

Prokaryotes are sensitive to their environment, and their genetic activity is controlled by specific proteins that interact directly with their DNA to quickly adjust to environmental changes. *Genetic expression* is the process where genotypes coded in the genes are exhibited by the phenotypes of the individuals. The DNA is copied by the RNA and then synthesized into protein. The process of transcription, which is the synthesis of RNA from a DNA template, is where the regulation of the gene expression is most likely to occur. The default setting for prokaryotes appears to allow for the continual synthesis of protein to occur, whereas in eukaryotes the system is normally off until activated.

An operon is a self-regulating series of genes that work in concert. An operon includes a special segment of genes that are regulators of the protein synthesis, but do not code for protein, called the promoter and operator. These segments overlap, and their interaction determines whether the process will start and when it will stop. RNA polymerase must create RNA by moving along the chromosome and "reading" the genes in the process of transcription.

RNA polymerase first attaches to the promoter segment, which signals the beginning of a particular DNA sequence. If not blocked, it passes over the operator and reaches the protein-producing genes where it creates the mRNA that instructs the ribosomes to create the desired protein. This process continues until the system is blocked by *repressor proteins.* Repressors bind with the operator and prevent RNA polymerase from proceeding to create mRNA by prohibiting access to the remainder of the protein-producing genes. As long as the repressor is binding with the operator, no proteins are made. However, when an *inducer* is present, it binds with the repressor, causing the repressor to change shape and release from the operator. When this happens, the RNA polymerase can proceed with transcription, and protein synthesis begins and continues until another repressor binds with the operator. Refer to the illustration *Transcription regulation.*

The *lac* operon model is probably the most studied and well known. In bacteria, such as E. coli, three genes are part of an operon that code for three separate enzymes needed for the breakdown of lactose, a simple sugar. A *regulatory gene,* located before the operon, continually makes repressor proteins that bind with the operator and prohibit the function of RNA polymerase. The system therefore remains off until a flood of lactose molecules binds with all available repressors and prevents their attachment to the operator. When the operator is free, the production of the enzyme to break down lactose continues until enough of the lactose molecules are broken down to then release repressors to recombine with the operator to stop production of the enzymes.

Transcription regulation.

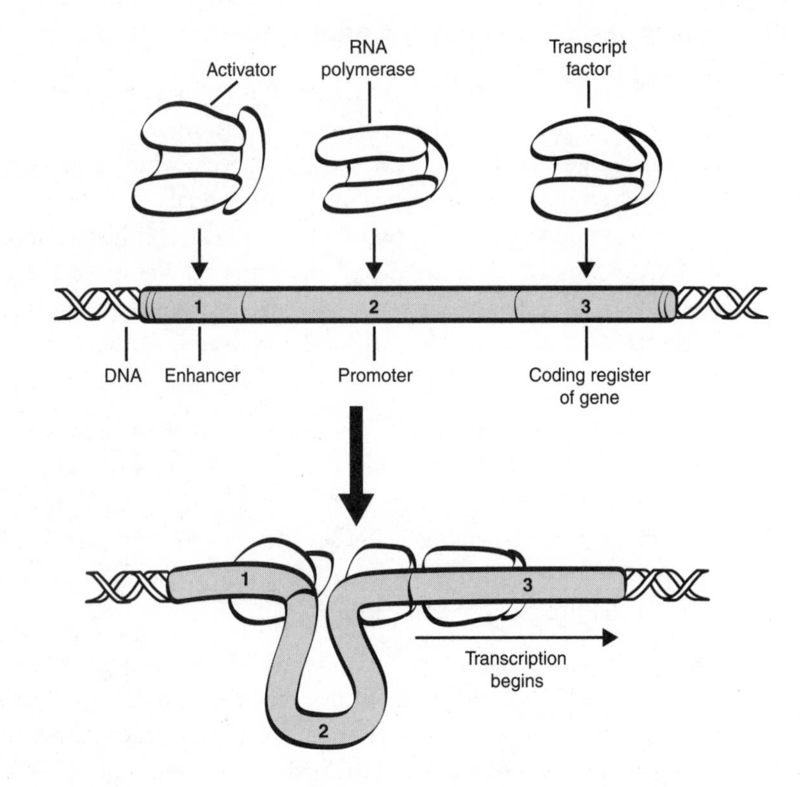

Two additional types of operons exist that operate in the same way except for the function of the operator. The *trp* operon differs because the repressor is active only when bonded to a specific molecule. For the remainder of the time, it remains unbonded and inactive in the absence of that molecule. Finally, in a positive twist, *activators* are used by a third type of operon to bond directly with the DNA, which allows the RNA polymerase to work more efficiently. Absent the activators, RNA polymerase proceeds at a slow rate.

Eukaryotes: Multiple Models of Gene Regulation

Unlike prokaryotes, multiple gene-regulating mechanisms operate in the nucleus before and after RNA transcription, and in the cytoplasm both before and after translation.

Histones are small proteins packed inside the molecular structure of the DNA double helix. Tight histone packing prevents RNA polymerase from contacting and transcribing the DNA. This type of overall control of protein synthesis is regulated by genes that control the packing density of histones. *X-chromosome inactivation* occurs when dense packing of the X chromosome in females totally prevents its function even in interphase. This type of inactivation is inherited and begins during embryonic development, where one of the X chromosomes is randomly packed, making it inactive for life.

Activator-enhancer complex is unique in eukaryotes because they normally have to be activated to begin protein synthesis, which requires the use of *transcription factors* and RNA polymerase. In general, the process of eukaryotic protein synthesis involves four steps:

1. Activators, a special type of transcription factor, bind to *enhancers*, which are discrete DNA units located at varying points along the chromosome.
2. The activator-enhancer complex bends the DNA molecule so that additional transcription factors have better access to bonding sites on the operator.
3. The bonding of additional transcription factors to the operator allows greater access by the RNA polymerase, which then begins the process of transcription.
4. *Silencers* are a type of repressor protein that blocks transcription at this point by bonding with particular DNA nucleotide sequences.

The processing and packaging of RNA both in the nucleus and cytoplasm provides two more opportunities for gene regulation to occur after transcription but before translation.

Adding extra nucleotides as a protective cap and tail to the RNA identifies the RNA as an mRNA by the ribosomes, and prevents degradation by cell enzymes as it moves from the nucleus into the cytoplasm.

RNA splicing occurs when "gaps" of nonprotein-code-carrying nucleotides called *interons* are removed from the code-carrying nucleotides, called *exons*, which are then connected to shorten the RNA molecule for conversion into tRNA and rRNA. The number of interons regulates the speed at which the RNA can be processed.

After the extra nucleotides have been added as a cap and tail and the RNA has been spliced, it moves to the cytoplasm where additional mechanisms of gene regulation exist.

The longevity of the individual mRNA molecule determines how many times it can be used and reused to create proteins. In eukaryotes, the mRNA tends to be stable, which means it can be used multiple times; which is efficient, but it prevents eukaryotes from making rapid response changes to environmental disruptions. The mRNA of prokaryotes is unstable, allowing for the creation of new mRNA, which has more opportunities to adjust for changing environmental conditions.

Inhibitory proteins prevent the translation of mRNA. They are made inactive when bonded with the substance for which they are trying to block production.

Post-translation control involves the selective cutting and breakdown of proteins that prevent the formation of the final product. In both cases, the hormone or enzyme required to finish or activate the final product may be rendered inactive.

Although much has been learned about inheritance since Mendel's time, the fundamentals remain the same. In sexual reproduction the offspring inherit one half of their genes from the father and one half from the mother. The chance of inheriting a particular gene can be estimated by pedigrees and Punnet squares. Every trait, feature, or characteristic is controlled by genes or a combination of genes. Numerous gene-regulating mechanisms activate and inactivate organism functions.

The Least You Need to Know

- Mendel's ideas are still useful in understanding heredity. They were based on prior scientific knowledge and his own meticulous experiments with garden peas.
- Meiosis makes haploid gametes; mitosis makes new cells. Both are subject to error, which greatly affects the offspring.
- There are many ways for mutations to occur.
- Eggs and sperm are made in the sex organs, but they are constructed differently.
- Prokaryotes regulate the action of genes differently than eukaryotes.
- Histones packed around the DNA in eukaryotes regulate gene expression.

Chapter 5

Genetic Engineering

In This Chapter

- How genetic engineering works
- Latest techniques for gene alteration
- Current useful products that have been created
- How "Dolly" was made
- Transgenic organisms
- The nonscience challenges to this type of research

This chapter will bring you up-to-date with the leading edge of scientific technology and its application in the real world. Recombinant technology is here today and will probably become the growth industry of the future as more and better techniques are discovered that have both a profit and a humanitarian incentive attached. This is it, the real stuff.

Of course with every change, whether good or bad, there are opposing points of view that weigh in to confuse the issue. Genetic engineering is one of these. There are scientific issues and legal issues that are being called down on both sides. You might want to learn these … in case you need a plasmid injection one of these days. Don't know what this is? Read on.

Regardless of where you are philosophically on these issues, they are with us today and will likely be around for a while. Some futurists predict that at a point in the future, random fertilization by consenting couples will cease to occur. It would be much less surprising to select the sperm and egg that will create the ideal offspring. Hmmmm, this could be interesting!

Fundamentals of Genetic Engineering

Genetic engineering is the process of transferring specific genes from the chromosome of one organism and transplanting them into the chromosome of another organism in such a way that they become a reproductive part of the new organism. The process that produces the resulting recombinant DNA involves four steps:

1. The desired DNA is cleaved from the donating chromosome by the action of *restriction enzymes*, which recognize and cut specific nucleotide segments, leaving a "sticky end" on both ends. The restriction enzymes also splice the receiving chromosome in a complementary location, again leaving "sticky ends" to receive the desired DNA.

Bionote

Restriction enzymes were discovered in the 1960s when they were observed cutting foreign DNA into small pieces as a means of protecting the host cell against virus intrusions.

2. The desired DNA fragment is inserted into a vector, usually a plasmid, for transfer to the receiving chromosome. Plasmids are an ideal vector because they replicate easily inside host bacteria and readily accept and transfer new genes. Plasmids are circular DNA molecules found in the cytoplasm of bacteria that bond with the desired DNA fragment with the help of the joining enzyme, *DNA ligase*, to create the resulting *recombinant DNA.*

3. When the host cell reproduces, the plasmids inside also reproduce, making multiple clones of their DNA. Because the plasmid DNA contains the desired as well as unwanted DNA clones, the entire product is referred to as a *gene library.* The desired gene is similar to one book in that library.

4. To recover the desired DNA, the current technology is to screen unwanted cells from the mixture and then use gel electrophoresis to separate the remaining genes by movement on an electric grid. Gel electrophoresis uses a positively charged grid to attract the negatively charged DNA fragments, thereby separating them by size, because the smaller ones will migrate the most. Radioactive or fluorescent probes are added, which attract and bind with the desired DNA to produce visible bands. Once isolated, the DNA is available for commercial use.

In 1973, researchers Cohen and Boyer created an interesting model for screening the host cells to finds the desired DNA fragment. In their experiment, they inserted the desired DNA and a DNA segment that made the host bacteria resistant to a particular antibiotic, tetracycline. When the antibiotic was applied to the general population, only those bacteria that had received the plasmid survived—so they knew their desired DNA fragment was located in the surviving bacteria.

Current Recombinant Cloning Technology

Reverse transcriptase is an enzyme that acts opposite of normal transcriptase. It uses RNA to code for DNA. It is also found in the virus linked to AIDS.

A more advanced method of producing DNA clones uses the enzyme *reverse transcriptase* and mRNA in a four-step process, which creates a pure segment of desired genes:

1. mRNA is made by a selected cell particularly for its genetic characteristics.
2. mRNA splices out the interons.
3. mRNA is isolated and used as a template with reverse transcriptase to make the complementary DNA.
4. The DNA product therefore contains only the desired DNA segment and the host cell will continue to produce the product.

With this type of emerging technology, the "shotgun" approach to cloning is simplified by not copying the entire genome of the individual, but only the specific genes required.

DNA Technology Applications

The use of *recombinant DNA* technology has become commonplace as new products from genetically altered plants, animals, and microbes have become available for human use. In 1997, Dolly made headlines as the first successfully *cloned* large mammal (sheep). Since then there have been many similar advances in medicine, such as treatments for cancer; many advances in agriculture, such as transgenic insect-resistant crops; and many advances in animal husbandry, such as growth hormones and *transgenic* animals (an animal that has received recombinant DNA).

Most biotechnologists envision DNA technological applications as one of the new frontiers in science with tremendous growth and discovery potential.

Medicine

Genetic engineering has resulted in a series of medical products. The first two commercially prepared products from recombinant DNA technology were insulin and human growth hormone, both of which were cultured in the E. coli bacteria. Since then a plethora of products have appeared on the market, including the following abbreviated list, all made in E. coli:

- **Tumor necrosis factor.** Treatment for certain tumor cells
- **Interleukin-2 (IL-2).** Cancer treatment, immune deficiency, and HIV infection treatment
- **Prourokinase.** Treatment for heart attacks
- **Taxol.** Treatment for ovarian cancer
- **Interferon.** Treatment for cancer and viral infections

In addition, a number of *vaccines* are now commercially prepared from recombinant hosts. At one time vaccines were made by denaturing the disease and then injecting it into humans with the hope that it would activate their immune system to fight future intrusions by that invader. Unfortunately, the patient sometimes still ended up with the disease.

Bionote

A **vaccine** is usually a harmless version of a bacterium or virus that is injected into an organism to activate the immune system to attack and destroy similar substances in the future.

With DNA technology, only the identifiable outside shell of the microorganism is needed, copied, and injected into a harmless host to create the vaccine. This method is likely to be much safer because the actual disease-causing microbe is not transferred to the host. The immune system is activated by specific proteins on the surface of the microorganism -e. DNA technology takes that into account and only utilizes identifying surface features for the vaccine. Currently vaccines for the hepatitis B virus, herpes type 2 viruses, and malaria are in development for trial use in the near future.

Agriculture

Crop plants have been and continue to be the focus of biotechnology as efforts are made to improve yield and profitability by improving crop resistance to insects and certain herbicides and delaying ripening (for better transport and spoilage resistance). The creation of a transgenic plant, one that has received genes from another organism, proved more difficult than animals. Unlike animals, finding a vector for plants proved to be difficult until the isolation of the *Ti plasmid*, harvested from a tumor-inducing (Ti) bacteria found in the soil. The plasmid is "shot" into a cell, where the plasmid readily attaches to the plant's DNA. Although successful in fruits and vegetables, the Ti plasmid has generated limited success in grain crops.

Creating a crop that is resistant to a specific herbicide proved to be a success because the herbicide eliminated weed competition from the crop plant. Researchers discovered herbicide-resistant bacteria, isolated the genes responsible for the condition, and "shot" them into a crop plant, which then proved to be resistant to that herbicide. Similarly, insect-resistant plants are becoming available as researchers discover bacterial enzymes that destroy or immobilize unwanted herbivores, and others that increase nitrogen fixation in the soil for use by plants.

Geneticists are on the threshold of a major agricultural breakthrough. All plants need nitrogen to grow. In fact, nitrogen is one of the three most important nutrients a plant requires (see Chapter 13). Although the atmosphere is approximately 78 percent nitrogen, it is in a form that is unusable to plants. However, a naturally occurring *rhizobium* bacterium is found in the soil and converts atmospheric nitrogen into a form usable by plants. These nitrogen-fixing bacteria are also found naturally occurring in the legumes of certain plants such as soybeans and peanuts. Because they contain these unusual bacteria, they can grow in nitrogen-deficient soil that prohibits the growth of other crop plants. Researchers hope that by isolating these bacteria, they can identify the DNA segment that codes for nitrogen fixation, remove the segment, and insert it into the DNA of a profitable cash crop! In so doing, the new transgenic crop plants could live in new fringe territories, which are areas normally not suitable for their growth, and grow in current locations without the addition of costly fertilizers!

Animal Husbandry

Neither the use of animal vaccines nor adding bovine growth hormones to cows to dramatically increase milk production can match the real excitement in animal husbandry: transgenic animals and clones.

Transgenic animals model advancements in DNA technology in their development. The mechanism for creating one can be described in three steps:

1. Healthy egg cells are removed from a female of the host animal and fertilized in the laboratory.

2. The desired gene from another species is identified, isolated, and cloned.

3. The cloned genes are injected directly into the eggs, which are then surgically implanted in the host female, where the embryo undergoes a normal development process.

It is hoped that this process will provide a cheap and rapid means of generating desired enzymes, other proteins, and increased production of meat, wool, and other animal products through common, natural functions.

Ever since 1997 when Dolly was cloned, research and experimentation to clone useful livestock has continued unceasingly. The attractiveness of cloning is the knowledge that the offspring will be genetically identical to the parent as in asexual reproduction. Four steps describe the general process:

1. A differentiated cell, one that has become specialized during development, with its diploid nucleus is removed from an animal to provide the DNA source for the clone.
2. An egg cell from a similar animal is recovered and the nucleus is removed, leaving only the cytoplasm and cytoplasm organelles.
3. The two egg cells are fused with an electric current to form a single diploid cell, which then begins normal cell division.
4. The developing embryo is placed in a surrogate mother, who then undergoes a normal pregnancy.

Human Disorders and Gene Therapy

Genetic disorders are the harmful effects on an individual caused by inherited genetic diseases or mutations. Usually genetic disorders are recessive, so they are only expressed in a small percentage of the population, but a much larger percentage are carriers. When expressed in the homozygous recessive individual, they often code for the wrong protein or amino acid sequence. There are many genetic disorders; however, two are common in today's population: hemophilia A and sickle-cell anemia.

Hemophilia A is a recessive sex-linked genetic disorder that is exhibited by approximately 1 in every 10,000 Caucasian males. Multiple genes code for the multistep process of blood clotting. Mutation in any one of them creates hemophilia A, the inability to form blood clots. Individuals with this disease must avoid all cuts and bruises, both internal and external. In severe cases, the individual may lose massive amounts of blood.

Sickle-cell anemia is a recessive genetic disorder that affects 1 in every 500 African Americans. A mutation of an allele causes the allele to code for a sickle-shaped hemoglobin molecule. The defective hemoglobin molecules do not transport as much oxygen as the hemoglobin in normal red blood cells because they tend to rupture. They also sometimes wedge in a blood vessel, blocking the flow of blood cells. Tissues and organs downstream from the obstruction may suffer serious damage. Interestingly, sickle cells are a survival advantage in certain areas because they are a defense against malaria and may protect some people from the disease.

Although most genetic disorders cannot be treated because of technology limitations, certain ones such as phenylketonuria (PKU) can be treated if discovered in time. For instance, a baby with PKU is maintained on a low-phenylalanine diet to prevent mental retardation caused by its buildup.

Most humans inherit genetic disorders because of the improper functioning of a particular gene sequence. In theory, replacing the defective gene with a healthy one should solve the problem, which is the essence of gene therapy. Although in its infancy as a treatment for disorders such as hemophilia and sickle-cell anemia, patients have received genetically engineered cells as an experimental treatment for missing genes. At this time the data are incomplete regarding the results. Currently, researchers are attempting to engineer cells, usually from bone marrow, to enhance the abilities of immune cells to fight off cancer and resist infection by HIV. This approach may lead to an effective treatment for nonhereditary diseases.

Human Genome Project

The Human Genome Project (HGP) is the most exciting breakthrough in human genetics in modern times! Geneticists from around the world collaborated to determine the nucleotide sequence for the complete human genome. This genetic map gives the location of each of the approximately 100,000 human genes composed of roughly 3 billion nucleotides.

The immense value of this knowledge will provide new understanding of how all genes work, how they are regulated, and how and why they create biological molecules. The human genome can then be compared to other known animal genomes to examine similarities and differences that may be useful in the creation of new genetic recombinations. Some of the knowledge gained may allow gene replacement and other gene-therapy strategies. For instance, it is known that sharks never contract cancer. If there is a cancer-inhibiting gene in sharks that could be incorporated into humans with no side effects, another serious health concern could be avoided. The possibilities open genetic engineering as a profitable, socially beneficial enterprise in the near future. It is estimated that there are more than 3,000 human genetic disorders!

Legal and Ethical Considerations

Many challenges to the new technology need to be addressed so that the research and treatment may proceed without violation of public trust and confidence.

Legal and ethical challenges can be classified into the following question categories:

- Who has the right to the cure?
- What will be the cost and availability?
- Do we have the right to alter a person's genes?
- Do we have the right to control the genetic complement of the human population and other eugenic considerations?

Several scientific questions also pose additional considerations:

- The development of new genes combinations increases genetic diversity, which is normally considered a positive effect, but may have unintended, unforeseen consequences.
- Creating new genes may also create new pathogenic organisms for which we have no cure.
- Do we have the ability to safely handle new genetically altered organisms?
- Certain bacteria have already been utilized to clean up oil spills; are there other uses?

The Least You Need to Know

- Clones are identical copies of DNA molecules, cells, or organisms that are genetically the same as their parent.
- Most genetic engineering techniques involve DNA segments, specialized enzymes, and often transmitting plasmids.
- Current cloning techniques are refined, specific, and efficient.
- Recombinant applications will continue to reform medical and agricultural methods as profitability powers future research.
- Genetic engineering is capable of transforming life—literally and figuratively.

Chapter 6

Genetic Variation and Natural Selection

In This Chapter

- Sexual reproduction and genetic variability
- How Darwin's theories account for speciation
- Why some traits are not explained by Mendel's theories
- Differences between sex-linked and sex-influenced genes

This chapter captures some of the ideas from the previous chapters and relates them in a whole new way; ways that even Mendel didn't encounter. But we will build upon Mendel's ideas to unlock the mystery of why some people who have red hair also have freckles and fair skin. We will also lay to rest why a colorblind mother always produces a colorblind son, but not necessarily a colorblind daughter.

Natural selection has a way of affecting the genetics of a population. Darwin understood it, and so will you. Again, simple concepts are put together to make bigger ones so that uneducated people will feel dumb. Thankfully, you will be on the enlightened side of the equation after reading this chapter.

Natural Selection

As discussed in Chapter 3, sexual reproduction and DNA mutation are the two primary processes that increase genetic variability. Although mutation is the only source of new alleles, the potential for a new allele combination is increased with every sexual reproduction from three primary processes:

- Random union of sperm and egg
- Crossing over during meiosis 1
- Independent assortment of homologous chromosomes

Although most genetic recombinations involving a recessive gene are biologically neutral, the potential for that gene to be expressed in the future still survives in that organism and therefore in that breeding population. Increasing genetic variation and the environmental effects on that variation form the basis for *natural selection*. Natural selection is a theory that states that those individuals who are best adapted to live in an area will survive and reproduce. Within a given population, there exists a normal degree of genetic variation that may or may not make an individual more adapted to the environment or, more importantly, changes in the environment. A species with a great deal of genetic variability is more likely to survive as a species in a changing environment than a species with limited variability.

Darwin and Natural Selection

The theories of Charles Darwin still form the foundation for our understanding of natural selection. It is still widely upheld that natural selection is based on five factors:

- Within a species, individual variations exist naturally. Some organisms are faster, more colorful, bigger, or smarter than others in their species. Individual variations sometimes are helpful, neutral, or harmful.
- More offspring are produced than can survive, and their success in their life struggle to gain food, shelter, and a mate determines their ability to successfully reproduce and pass on their genetic complement.
- The reproductive rates of individuals are unequal, or favored based on environmental pressures.
- Environmental conditions determine the reproductive success of certain individuals because they possess a trait that allows them an advantage in that environment.

- Individuals who are able to provide the necessary food, shelter, and avoid predation reproduce more successfully than others. Over generations, the characteristics of the population change as those individuals more successful in reproduction populate the species. Less-successful types do not pass on as many traits because they have fewer descendants. Over time, failure to reproduce at a rate equal to or greater than the mortality rate leads to extinction. The process whereby the frequency of certain traits within a species change by uneven reproduction rates caused by natural selection is called *evolution.*

Bionote

Darwin wrote extensively about the Galapagos finches, which he studied while on a research voyage on the HMS *Beagle.* He carefully observed and noted that the variation in the beak structures of the finches had created certain advantages or distinctions in their quest for food. For instance, some beaks were designed for crushing seed shells, whereas others for catching insects. He is most closely associated with his study of finches.

In Darwin's model, natural selection did not necessarily prohibit an individual from reproducing, but rather favored the most adapted individuals with a greater chance for reproduction, the possibility of greater birth numbers, and the increased development and survival of the offspring. So natural selection allows the total number of certain members of a population to decline or become extinct because of environmental pressure, while others increase accordingly.

The classic case study of natural selection was recorded by H. B. D. Kettlewell in 1952. In a most interesting experiment, he concluded that an environmental influence, *predation* by birds, affected the total number of reproducing moths depending on their color. Kettlewell was working at Oxford University at the time and discovered a shard of information from the 1840s, 100 years earlier, that noted the first appearance of a dark morph of the peppered moth. Until that time, only white or pepper-colored moths had been observed. He connected the dates with the onset of heavy industrial output in that area. He also knew that the factories at that time and in his time produced voluminous daily clouds of black smoke heavily laden with soot. He also knew that the peppered moths were common all over England; they were nocturnal and hid on tree trunks during the day; they were preyed upon by many species of birds. He hypothesized that the birds

Bioterms

Predation is a relationship between species where one species, the predator, consumes the other species, the prey.

were preying upon the moths that were less camouflaged and therefore easier to see. In so doing, they were favoring the one morph type over another, which created uneven reproductive rates that favored an increase in one type of moth over the others.

In his experiment, he released light-colored moths in soot-covered forests and an equal number of dark moths in normal forests. He tagged each moth with a touch of paint and then set traps for their recapture. In the soot-covered dark forest, he recaptured mostly dark moths; in the lighter forest, mostly light moths. He concluded the uneven recapture rates were based on uneven predatory rates. To confirm his suspicion, he observed and took pictures of birds preying upon moths whose body color contrasted with the environment while apparently overlooking the more camouflaged moth. Those moths best adapted to the environment survived and reproduced; the others did not. Kettlewell's experiment is an example of how environmental pressures can determine the characteristics of species.

Bionote

Alfred Wallace also proposed a theory of natural selection at the same time as Darwin; however, Darwin's name is linked with the idea probably because of increased awareness provided by his book *Origin of Species*, published in 1859.

At this time in history, Darwin's theories met great opposition from other scientists and religious leaders because of their newness and related controversial nature. Much of the criticism appeared to stem from a misunderstanding of his ideas.

Post-Mendelian Inheritance Factors

Recall from Chapter 4 that most animals, including you, are diploid, meaning that each of your traits is controlled by the interaction of at least two genes. Mendel's work concentrated on the effect of a single gene from the mother and a single gene from the father to determine the genotype and phenotype of the offspring based on dominant and recessive genes. We now know that often multiple genes are involved in strange and intriguing ways to influence the phenotype of the offspring. The next section focuses on seven such mechanisms.

Bionote

More than 200 human traits are controlled by a single gene pair, such as dwarfism, cataracts, cystic fibrosis, and albinism.

Incomplete Dominance

Whenever two genes combine and the offspring's phenotype is a compromise between the effects of the two genes, then neither gene has expressed dominance over the other gene. In fact, one gene is incompletely dominant over the other gene that is inactive.

Mendel's work with peas never encountered incomplete dominance because the heterozygous genotype always expressed the dominant gene in the phenotype.

The active gene is apparently not strong enough to compensate for the loss of effect by the inactive allele. The classic example of *incomplete dominance* is the cross between a red snapdragon and a white snapdragon. According to Mendel, the offspring should be either red or white. In reality they are all pink, indicating that neither allele is completely dominant but one is incompletely dominant. Therefore, a weak expression of one allele, the red one, is actually expressed in a heterozygous offspring.

Incomplete dominance becomes evident in the F2 generation as detailed in the following Punnet square analysis of a fictitious trait: B = blue, b = yellow. A cross between a blue (BB) parent and a yellow (bb) parent, BB × bb, should give all blue offspring in the F1 generation. Refer to the illustration *Parents: BB × bb.*

F1

	b	b
B	Bb	Bb
B	Bb	Bb

Parents: BB × bb.

Instead, the Bb genotype is green, a blending of the two colors. This indicates that the B gene is not strong enough to dominate the effect of the b allele completely. If two heterozygous individuals are mated, the results are even more intriguing. Refer to the illustration *Parents: Bb × Bb.*

F2

	B	b
B	BB	Bb
b	Bb	bb

Parents: Bb × Bb.

The genotypes are BB (1); Bb (2); bb (1). A 1:2:1 ratio is typical of Mendelian inheritance patterns. However, the phenotypic ratio is: blue (1), green (2), yellow (1). This 1:2:1 ratio is not expected by Mendel's predictions.

Codominance

Codominance is a type of allele dominance that occurs when both alleles are active and expressed, as sometimes happens in a heterozygous individual. This phenomenon is most commonly displayed in the roan coat color of horses and cows. A roan color is actually the result of a codominance by a gene for red hair and a gene for white hair. In a roan-colored animal, there are equal numbers of both colors evenly dispersed within the coat, creating a ghostly brown color. Another example is Erminette chickens. They are a mottled black and white color because they have roughly equal numbers of black and white feathers. In their case, a gene for white feathers and a gene for black feathers are codominant in the heterozygous individuals.

Multiple Alleles

Although individual humans carry a maximum of two alleles that control the expression of a trait, often there are more than two different alleles in a given population that code for the same trait. When three or more alleles are available for a particular gene, the inheritance of those alleles follows a *multiple alleles* pattern. The most common case for demonstrating multiple alleles is the inheritance of human blood types: A, B, AB, and O. Combinations of three possible alleles give rise to the four blood types. Because you can only have two alleles for any gene, not everyone carries every type of allele available; but within a population, all of the multiple alleles are present. Sexual reproduction within that population keeps the alleles available for future generations and continues the genetic variation.

An expanded Punnet square shows the possible blood type combinations that exist in the human population. Refer to the illustration *Human population blood type combinations.*

Remember although there are three possible alleles listed on each axis, a human can only possess two of these. However, other individuals in the human population may carry a different set of two alleles, so that all three types are present and active in the human population.

Human population blood type combinations.

	A	B	O
A	AA	AB	AO
B	AB	BB	BO
O	AO	BO	OO

Key: AA = A; AB = AB; BB = B; OO = O
AO = A BO = B

Linked Genes

Each chromosome normally carries thousands of genes. When certain neighboring genes on the same chromosome are inherited together, they are considered *linked genes.* Linked genes are an anomaly in Mendel's theory of independent assortment because they tend to remain together as a segment and therefore function as a cluster of genes rather than individual genes. In humans, the autosomal genes for red hair, freckles, and a fair complexion are linked genes.

Sex-Linked, or X-Linked Genes

The linkage in *sex-linked traits* is completely different from linked genes. A sex-linked trait is one whose allele is located on one of the sex chromosomes, usually the X for autosomal traits. Most sex-linked genes are recessive and are only expressed in homozygous females and males. Because males normally inherit one X chromosome from the mother only, the recessive traits found on the X chromosome are passed from mother to son. It is more difficult to pass an X-linked trait to a daughter because she inherits an X from both parents, so both parents would have to possess and pass on the recessive gene. The recessive gene on the X chromosome is expressed in males because it is the only gene present. In females, it can be hidden by the presence of a dominant gene on the other X chromosome. In humans, sex-linked traits include colorblindness, albinisms, and hemophilia.

Refer to the illustration *Colorblindness pedigree.* A heterozygous "carrier" mates with a "carrier" female (X^C = gene for colorblindness). A carrier is an individual who is heterozygous for a condition but does not display the trait. Therefore, a carrier for colorblindness is not colorblind but can pass on the gene for colorblindness.

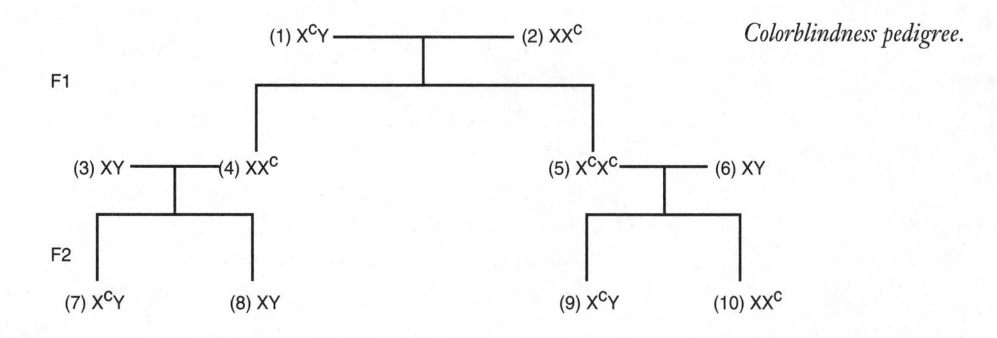

Colorblindness pedigree.

In the F1 generation, one colorblind female and one carrier female each mate with a noncolorblind male. In the F2 generation, individuals 7 and 8 are brothers, one is colorblind, the other is not. The colorblind male inherited the colorblind gene from the heterozygous carrier mother; the noncolorblind son inherited the normal X chromosome that did not contain the gene for colorblindness. Likewise, individual 9 could only receive a gene for colorblindness from his mother, number 5. The sister, number 10, is a heterozygous carrier.

Sex-Influenced Genes

Certain genes are activated by hormones secreted by the sex organs, so they display different phenotypes when found in males than they do when found in females, even though their genotypes are identical. The standard sex-influenced study in humans is male pattern baldness, which is influenced by testosterone and can affect both males and females. However, it is dominant in males and recessive in females! In the homozygous state, both males and females become bald; in the heterozygous genotype, females show no signs of baldness, but males become bald. Interestingly, sex-influenced genes are generally not found on the sex chromosomes, so both sexes can have the same genotype and different phenotypes!

Polygenic Inheritance

So far we have limited gene expression to those traits expressed by the action of one gene. In most primates, including you, traits are governed by the interaction of multiple genes. The genes of *polygenic traits* may be interspersed on the same chromosome or on completely different chromosomes. The additive effects of numerous genes on a single phenotype create a continuum of possible outcomes. A typical example is eye color in humans, which under normal genotypes is in varying shades of ebony brown to crystal blue, to Kelly green, and all points in between. Polygenic traits are also most susceptible to environmental influences. For instance, tallness is controlled by polygenes for skeleton height, but their effect may be retarded by malnutrition, injury, and disease.

Bionote

Polygenic inheritance is the opposite of *pleiotropic* inheritance, in which a single gene affects several characteristics.

Environmental Influences

Natural environmental influences on the expression of genotypes tend to be an advantage for the individual, whereas man-made environmental influences are both positive and negative for the individual. Natural environmental influences include the phenomenon of color change in the Arctic fox from red-brown in the summer months to pure white during the winter season for better camouflage. The genes that produce the red-brown summer pigment are blocked by cold temperatures, causing the hair to grow with no color (therefore, white). Another colorful example is the interaction between the color of the hydrangea flower, which is blue in acidic soils and pink in alkaline soils. A recent study also linked improved diet in infants and adolescents to a taller average height in the United States and Europe with the opposite effect in famine-stricken populations. The genetic complement of an individual is inherited; however, the environmental effect on these genes may alter their application and expression.

The Least You Need to Know

- Genetic variation in a species is desirable.
- Mutation and sexual reproduction increase genetic variation.
- Darwin's theory of natural selection explains how natural variation exists within a population and is acted upon by forces in the environment to create unequal reproductive rates within a species.
- Many inheritance patterns cannot be explained by Mendelian genetics, but are typical in human genetic patterns.
- All characteristics in every living organism are coded for by DNA, but environmental and man-made influences may limit the full expression of the DNA blueprint.

Part 3 Evolution, Natural Selection, and Speciation

People have historically attempted to solve the question of how species change over time. Evolutionary topics are presented in genetic-environmental interactive terms, which show cause and effect simultaneously for better understanding. Supporting data are provided through radiometric analysis of fossil evidence that also generates a geologic time scale. You may not be the only person to be confused by science; the chapter on systematics clarifies the confusion and chaos that surrounds the classification system for all living things.

Chapter 7

Historical Development and Mechanisms of Evolution and Natural Selection

In This Chapter

- Darwin's evidence and experimental data
- Darwin's theory of natural selection and evolution
- The Malthusian dilemma
- How the mechanisms of selection affect a population

Did a giraffe's neck grow to great lengths for the animal to survive or because organisms with increasingly longer necks reproduced more effectively and, hence, had a greater chance of passing on their genes than the short-necked variety? This is an interesting question with many answers; you will know the correct one soon.

This chapter may save you the trouble of stretching to become taller; it will also present you with one of the great dilemmas that is still with us. Hopefully you can solve it. Darwin's theories are presented in a historical context so you can develop the sense of understanding in incremental, sequential units that may stay with you for a while, at least through the next test. You will also be able to visualize the effects of natural selection on population trends. This is good stuff!

Early Theories of Evolution

The scientific community has always displayed an interest in the understanding of how life began. A rekindled interest emerged in the 1800s as new geologic and biologic discoveries added to the available knowledge.

Lamarck

Jean Baptiste Lamarck, a French scientist, proposed that species change because of the use or disuse of features, such as tails or arms. He purported the "use it or lose it" idea. He believed that excessive use of a feature would cause it to grow and lack of use would invite atrophy. He also stated that an individual's increased or decreased features could be passed on to the individual's offspring.

Lamarck hypothesized that *acquired traits*, not genes, were passed on to offspring. The Lamarckian concept of evolution appeared 50 years before Darwin. Unfortunately, he did not possess an understanding of gene behavior and heredity. However, his ideas started the discussion of evolution in the scientific community and, combined with the three additional ideas that follow, serve as a starting point for Darwin and others.

- Populations change over time in response to environmental pressure (same as Darwin) and may have evolved from small numbers of starting organisms.
- Organisms could actively try to change to be more adapted to the environment and then pass that change on to their offspring. For instance, the ancestors of fish saw the advantage of being able to swim faster, so over generations they developed fins instead of legs to increase their speed in water, thereby making them more efficient at getting food and eluding predators. Because this change helped and therefore was desirable to the organism, it was passed on to all offspring.
- The "use it or lose it" concept began with Lamarck, who believed that organisms could reshape their bodies in response to a desire to change. Appendages that were no longer needed would not be conveyed to the next generation, whereas other desirable changes could be added.

Bionote

According to Lamarck, giraffes grew long necks because they wanted to reach leaves higher up on the tree.

Many period biologists agreed with Lamarck that these *acquired characteristics* were inheritable factors. However, with no understanding of genetics, they could not explain how these things could happen.

Lyell

Most people in Darwin's time assumed that the earth was relatively young and that mountains, rivers, and other surface features did not change over time, but were formed by a catastrophic event. For instance, lakes might have formed when a meteorite collided with the earth and then rain filled the resulting crater with water.

Several geologists proposed contrary theories stating that the earth is likely at least four billion years old and that surface features normally change slowly over time. In 1788, James Hutton proposed that the mountains and other surface features gradually change, which supported the hypothesis that the earth was much older than the prevailing thought. Charles Lyell expanded on Hutton's work when he added the concept of *uniformitarianism*, which stated that the events and processes that created the mountains, lakes, and other surface features are still happening and an understanding of these events was required to explain the formation of existing features. Lyell also proposed that the earth itself had internal movements, as evidenced by earthquakes, that provided further opportunities for change. As scientists began to find and study fossils, Lyell's explanations gained credibility. Lyell contributed to Darwin's thinking in two ways:

- Past events must be explained in terms of today's events.
- The earth's history provides ample geologic time for slow change.

Artificial Selection

Darwin realized that a preferred method of selection had been taking place on a local level for a long time. In his conversations with local farmers, he learned that variability exists in every population and was the basis for enhancing yield and productivity. For instance, in a given wheat field, some plants developed greater seeds than others; likewise, some cows gave more milk than others. Those individuals that contributed the most, the farmers allowed to reproduce; all others were prohibited. In so doing, the farmers were attempting to increase their productivity, and they also selectively favored certain organisms for reproduction. Darwin reasoned that in this form of *artificial selection*, man could drastically change the appearance of a species in a relatively short period of time from the original wheat or cow to their descendants who were more capable of producing desirable results than their ancestors.

Malthusian Dilemma

Thomas Malthus was actually an English clergyman who worried about the distribution of wealth and resources. His observations and reasoning became known as the Malthusian dilemma, which states that populations multiply geometrically while food sources multiply arithmetically. If left to follow course, the human population would eventually grow larger than the ability of the environment to feed everyone. He predicted large mass starvations that crossed international borders with the possibility of resource wars, famine, and plague. Oddly enough, wars, famine, and plague *were* the answer to the dilemma, because they served to reduce the reproductive population, an idea that did not escape Darwin. He expanded the phenomenon to all creatures, not just humans, and understood that more offspring are born than will survive to reproduce. Refer to the illustration *The Malthusian dilemma.*

The Malthusian dilemma.

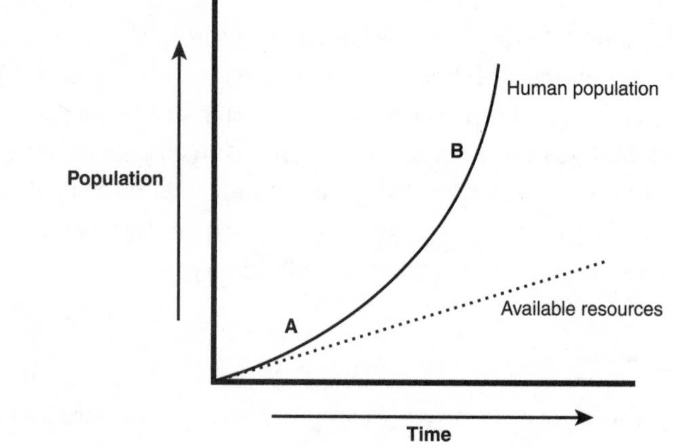

At point A on the graph, the amount of available resources is adequate to sustain a stable or slightly growing population. At point B, the growth of the human population has reached exponential rates, but the growth of available resources has not increased as dramatically. As a result, Malthus predicted that the intense competition for the resources would lead to war, famine, and plagues.

Natural Selection and Evolution

Darwin had the advantage of data from previous experimental findings, which he studied and compiled into his theories. It should also be noted that Darwin had access to these studies even though they were often in foreign countries and many years before his birth. Darwin also was aware of the extreme criticism that previous theories of evolution brought from the politicians and religious leaders of the time. To prevent a personal attack on his controversial ideas, he withheld his manuscript for 14 years.

Darwin's Theories

Darwin's first theory on evolution, called *descent with modification*, was based on a combination of Lamarckian ideas and recent fossil discoveries. He theorized that organisms living today have been changed over time and probably stemmed from a single, or perhaps a few, starting ancestral organisms. This also explained why the members of a species living in one area may appear different from a similar species living in a different environment. While in South America, he examined fossilized armadillos from a local archaeological dig and noted that although similar to modern armadillos, they had minute differences. Darwin saw descent with modification as evolution.

Darwin's second theory, called *modification by natural selection*, expands on Malthus's work by stating that although populations have the ability to grow uncontrollably, they do not do so for very long because of natural limiting factors such as increasing death rate or decreasing birth rate. He further referenced the farmers' understanding by stating that within a population, individual variability exists and may provide an advantage for one individual over another. Individuals who possess more desirable traits are more *adapted* to their environment, so they are able to reproduce more successfully and populate succeeding generations with a higher percentage of descendants than those individuals less adapted to who environment. Likewise, whether artificial or natural, the possession of certain traits that are favored provides the individual with an *adaptive advantage*.

Variation

The process of selection, both artificial and natural, favors those traits that are considered desirable as well as those traits most adapted to their environment at the expense of those who are not. In so doing, selection tends to reduce the genetic variability of the population as the unequal reproductive rates favor the descendants of one individual over another, leading to decreased genetic variability. In certain haploid organisms, the lack of genetic variability is often a reason for extinction or greatly reduced numbers.

Fortunately, in diploid organisms, populations are less likely to become genetically similar because recessive genes are hidden in a heterozygous individual. Because of this *heterozygous advantage*, a recessive allele can remain viable and present in a population, because its phenotype is never expressed and therefore never selected for or against. Albinism is a condition characterized by organisms with a lack of pigment. It is a recessive gene that is carried in both plants and animals, including the human population, but is not expressed unless the offspring inherits the recessive gene from both parents.

Likewise, populations on the endangered-species list may have populations so small that their genetic variability is subsequently low. Adding to this problem is the unlikely chance that new reproductive members will be found to increase the genetic diversity or spontaneous random mutations creating new alleles to increase genetic variability. In keeping with "survival of the fittest," the most fit individual is the one most capable of passing on its genes, so the chance for the survival of the genes is critical in small populations, such as endangered species.

Natural-Selection Mechanisms

Although the genotype of a population is affected by natural selection, it is the environment's effect on the individual phenotype that defines "fittest."

Polymorphism and Polygenic Traits

Individual genetic variations must also function within the existing genome for polygenic traits and related characteristics to function correctly. Polygenic inheritance provides a continuum of trait expression such as hair color, even to the extent that certain species are *polymorphic*—having two or more dominant phenotypes that are both commonly expressed in nature. In the case of polymorphic moths, a dark environment favors the black variant over the white, which is easily discovered and consumed by predators. Both polygenic traits and polymorphism create phenotypic variations that are acted upon differently by selection, as described in the next sections.

Directional Selection

Directional selection works by selecting one phenotypic extreme at the cost of excluding the other phenotypic extreme. This allows the offspring of the favored extreme to reproduce more effectively and therefore dominate the population. As a result, the phenotype of successive generations of offspring move in a definite direction because of natural selection. It is most common when a population colonizes a new territory, or when new environmental changes such as an herbicide application are introduced into the system.

Directional selection explains why certain diseases are now resistant to long-standing medicines. Interestingly, it is not because the individual germs grew an immunity or resistance to the medicine (as Lamarck would say), but that as the medicine is applied over the years to a large population, sooner or later the natural variability in the germ population will produce a germ that is unaffected by the medicine. So while all the other germs die and therefore do not reproduce, the naturally resistant germ lives and continues to reproduce until an entire population of resistant germs is formed. The phenotype for that germ has moved in a direction that was defined by natural selection operating on and favoring an extreme phenotype.

Directional selection can be demonstrated by analyzing the following graphs. In the illustration *Graph 1*, the bell-shaped curve typifies the normal variation found within a species. The shaded area exemplifies those organisms favored for reproduction by natural selection.

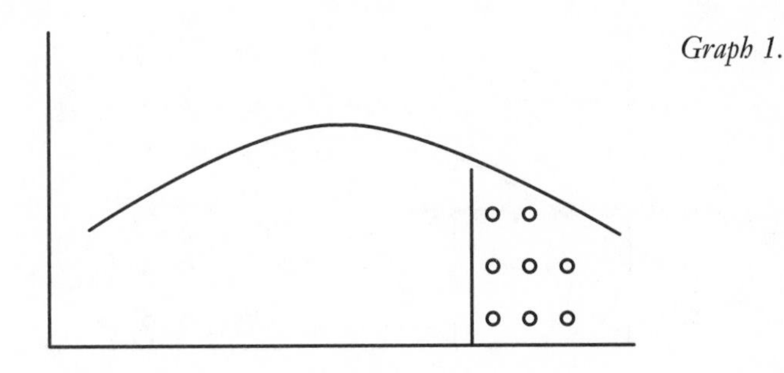

Graph 1.

In the illustration *Graph 2*, the population is the same after many generations. The shaded area indicates that the entire population is a descendant of the extreme phenotype found in Graph 1. The phenotype of the entire population has changed because of the directional selection forced by natural selection.

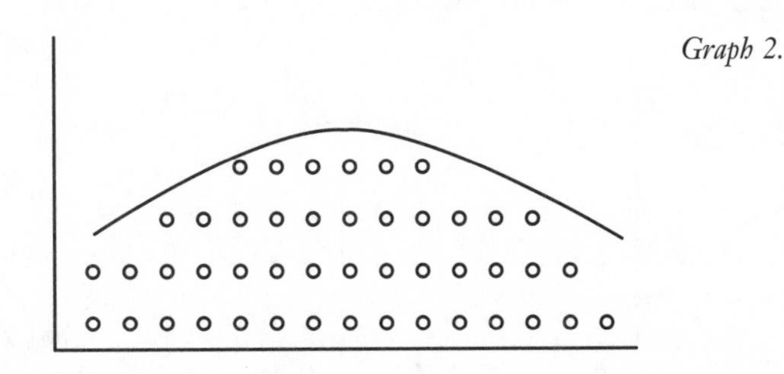

Graph 2.

Diversifying Selection

Diversifying selection favors both phenotypic extremes at the expense of intermediate phenotypic ranges. From Chapter 6, remember that black-and-white moths are more camouflaged in backgrounds of sharp contrasting colors than the intermediate gray type, thus making the black-and-white type more likely to escape predation and reproduce in greater numbers. Diversifying selection is the opposite of *stabilizing selection.*

Diversifying selection forms multiple phenotypes that exhibit successful breeding seasons within a population. The result is the prevalence of two (or more) morphs within a population. Examine the illustration *Graph 3*. Again the bell-shaped curve represents the normal variation found in all populations. The shaded areas represent the two different phenotypic extremes favored by natural selection.

Graph 3.

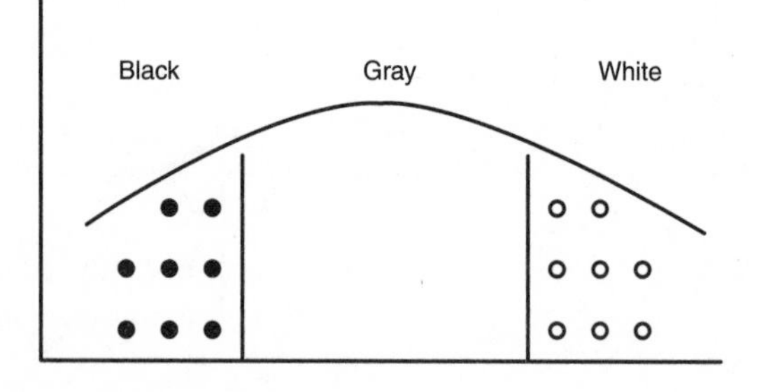

The illustration *Graph 4* represents the population after many generations. The intermediate, or gray, phenotype has failed to maintain reproductive vigor, so that species has been eliminated from the population. The remaining are descendants of the phenotypic extremes in Graph 3. Graph 4 demonstrates the phenotype diversification of a population by natural selection.

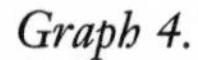
Graph 4.

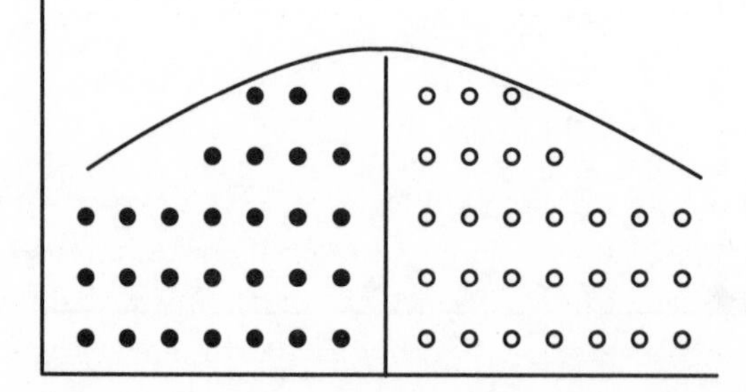

Stabilizing Selection

The opposite of diversifying selection is *stabilizing selection* because it favors the intermediate phenotype at the expense of the extremes. Whereas a diversifying mechanism signals a changing environment, stabilizing selection indicates a stable environment that tends to reduce phenotypic variation.

Stabilizing selection favors the intermediate phenotypic range at the exclusion of the extremes. The illustration *Graph 5* is similar to Graph 3, except natural selection favors the nonshaded, bulk of the population.

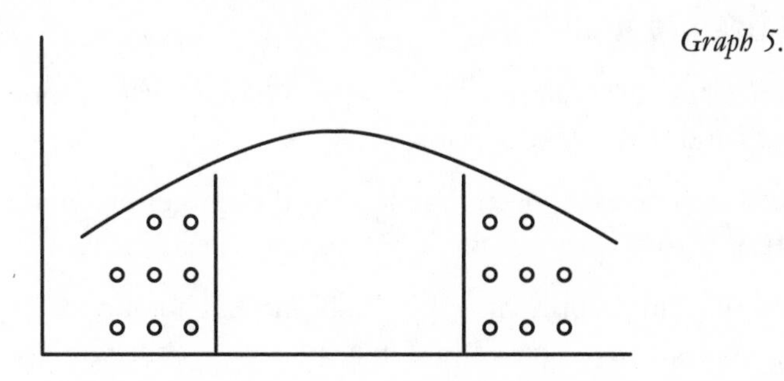

Graph 5.

The illustration *Graph 6* demonstrates the effect of stabilizing selection. The phenotype extremes have been eliminated. The result is a more homogeneous population.

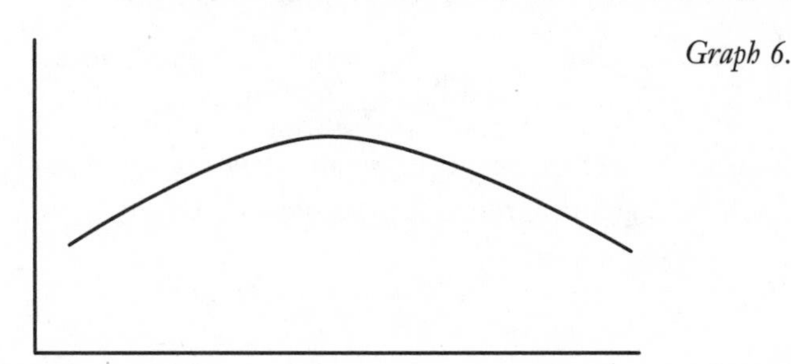

Graph 6.

Sexual Selection

The preference of one mate over another is *sexual selection.* In some cases, a demonstrative mating dance, elegant plumage, or overall size and physical/mental characteristics may influence the decision to mate or not mate with a given individual. In sharp contrast to evasive and camouflaged behaviors to avoid predation, these acts often appear ostentatious, regardless of the negative possible consequences, such as increased predation. Some will mate and pass on their genes; others will not. The results of this type of selection are most closely aligned with directional or diversifying selection as it affects the overall population structure. Sexual selection is different in that the form of selection is generated by factors within other members of the same species as opposed to an outside event.

The favoring of one phenotype over another by the biotic and abiotic factors in an environment create unequal reproductive rates for individuals in that population. Survival of the fittest essentially means who can reproduce in greater numbers. Reproductive success is the essence of natural selection.

The Least You Need to Know

- Evolution has always been a controversial topic. Early proponents were ridiculed and their ideas minimized.
- Artificial selection had been practiced in the agricultural and animal husbandry industries before Darwin's theory of natural selection was released.
- Natural selection represents multiple biotic and abiotic forces that create an uneven reproductive success rate in a species that changes the characteristics of that species as generations continue to reproduce the preferred characteristics.
- The modern theories of natural selection and evolution were derived from the experiments of many scientists working in different areas.
- Genetic variation allows a species but not necessarily an individual or groups of individuals to survive environmental changes.
- Many selection mechanisms increase and decrease genetic diversity within a population by regulating reproductive success.

Chapter 8

Microevolution and Macroevolution

In This Chapter

- Distinguishing microevolution from macroevolution
- Factors that affect genetic equilibrium
- What causes new species to be formed
- How genetic drift and gene migration affect allele frequency
- Establishing the age of fossils
- Determining the rate of speciation

The concepts covered in Chapter 7 are carried forward to higher levels of understanding in this chapter. You may want to review Chapter 7 before moving forward. Although this chapter is not complicated, it may save you time if you refresh your memory.

The selection strategies caused by natural forces create interesting patterns in the genetics of the affected populations. These patterns are easily described and understood, trust me. An extension of these patterns also describes how new species are formed. As simple concepts begin to build on themselves, it is important to remember that it all fits together in a nice package. It does make sense, but, as "they" say, you may not know it's a horse if you only have access to the tail.

Microevolution

Microevolution is the change in the genome, or gene pool, for a given species in a relatively short period of geologic time by the alterations of successfully reproducing individuals within a population. Evolution is the change in the gene pool of a species such that the organism becomes a new species. Environmental pressure favors changes that allow an organism to reproduce more successfully in that environment. Some environmental conditions are more harsh than others, and organisms may have to adapt more to survive—areas where the environmental pressures are stable, or the organisms have adapted to it, exhibit nonevolving populations. Interestingly, in a nonevolving population, the *allele frequency*, *genotype frequency*, and *phenotype frequency* remain in *genetic equilibrium*. In other words, the random assortment of genes during sexual reproduction does not alter the genetic makeup of the *gene pool* for that population. This phenomenon was illuminated by the mathematical reasoning of a German physician, Weinberg, and a British mathematician, Hardy, both working independently in 1908. Their combined efforts are now known as the Hardy-Weinberg equilibrium model.

Hardy-Weinberg Equilibrium

To demonstrate the Hardy-Weinberg equilibrium, assume G and g are the dominant and recessive alleles for a trait where GG = green, gg = yellow, and Gg = orange. In our imaginary population of 1,000 individuals, assume that 600 have the GG genotype, 300 are Gg, and 100 are gg. The allele and genotype frequency for each allele is calculated by dividing the total population into the number for each genotype:

GG = 600/1,000 = .6

Gg = 300/1,000 = .3

gg = 100/1,000 = .1

After the allele frequency has been determined, we can predict the frequency of the allele in the first generation of offspring.

First, determine the total number of alleles possible in the first generation. In this imaginary case, because each organism has 2 alleles and there are 1,000 organisms, the number of possible alleles in the first generation of offspring is:

$2 \times 1{,}000 = 2{,}000$

Next, examine the possibility of each allele. For the G allele, both GG and Gg individuals must be considered. Taken separately,

GG = $2 \times 600 = 1{,}200$

+ Gg = 300

1,500

The frequency for the G allele is therefore:

$1{,}500/2{,}000 = .75$

For the g allele, the calculation is similar:

Gg = 300

+ gg = $2 \times 100 = 200$

500

The frequency for the g allele is therefore:

$500/2{,}000 = .25$

Bionote

The letter p is used to identify the allele frequency for the dominant allele (.75) and q for the recessive allele (.25). Note that $p + q = 1$.

Hardy-Weinberg can also predict second-generation genotype frequencies. From the previous example, the allele frequencies for the only possible alleles are $p = .75$(G) and $q = .25$(g) after meiosis. Therefore, the probability of a GG offspring is $p \times p = p^2$ or $(.75) \times (.75) = 55$ percent. For the gg possibility, the allele frequencies are $q \times q$ or $(.25) \times (.25) = 6$ percent. For the heterozygous genotype, the dominant allele can come from either parent, so there are two possibilities: Gg = $2pq = 2(.75)(.25) = 39$ percent.

Note that the percentages equal 100, and the allele frequencies (p and q) are identical to the genotype frequency in the first generation! Because there is no variation in this hypothetical situation, it is in Hardy-Weinberg equilibrium, and both the gene and allele frequencies will remain unchanged until acted upon by an outside force(s). Therefore, the population is in a stable equilibrium with no innate change in phenotypic characteristics. As mentioned in Chapter 7, populations tend not to stay in Hardy-Weinberg equilibrium for very long because of environmental pressures.

The Hardy-Weinberg equation highlights the fact that sexual reproduction does not alter the allele frequencies in a gene pool. It also helps identify a genetic equilibrium in a population that seldom exists in a natural setting because five factors impact the Hardy-Weinberg equilibrium and create their own method for *microevolution.* Note that the first four do not involve natural selection:

- Mutation
- Gene migration
- Genetic drift
- Nonrandom mating
- Natural selection

Refer to Chapter 7 for a more in-depth coverage of the natural selection model.

Mutation

A primary mechanism for microevolution is the formation of new alleles by *mutation.* Spontaneous errors in the replication of DNA create new alleles instantly while physical and chemical mutagens, such as ultraviolet light, create mutations constantly at a lower rate. Mutations affect the genetic equilibrium by altering the DNA, thus creating new alleles that may then become part of the reproductive gene pool for a population. When a new allele creates an advantage for the offspring, the number of individuals with the new allele may increase dramatically through successive generations. This phenomenon is not caused by the mutation somehow overmanufacturing the allele, but by the successful reproduction of individuals who possess the new allele. Because mutations are the only process that creates new alleles, it is the only mechanism that ultimately increases genetic variation.

Bioterms

A **mutation** is an inheritable change of a gene by one of several different mechanisms that alter the DNA sequencing of an existing allele to create a new allele for that gene.

Gene Migration

Gene migration is the movement of alleles into or out of a population either by the immigration or emigration by individuals or groups. When genes flow from one population to another, that flow may increase the genetic variation for the individual populations, but it decreases the genetic variability between the populations, making them more homogeneous. Gene migration is the opposite effect of *reproductive isolation,* which tends to be genetically near the Hardy-Weinberg equilibrium.

Genetic Drift

Genetic drift is the phenomenon whereby chance or random events change the allele frequencies in a population. Genetic drift has a tremendous effect on small populations where the gene pool is so small that minor chance events greatly influence the Hardy-Weinberg arithmetic. The failure of a single organism or small groups of organisms to reproduce creates a large genetic drift in a small population because of the loss of genes that were not conveyed to the next generation. Conversely, large populations, statistically defined as greater than 100 reproducing individuals, are proportionally less affected by isolated random events and retain more stable allele frequency with low genetic drift.

Two types of genetic drift act when a large population is modified to be considered statistically as a small population: fragmentation effect and pioneer effect.

The *fragmentation effect* is a type of genetic drift that occurs when a natural occurrence, such as a fire or hurricane, or man-made event, such as habitat destruction or overhunting, unselectively divides or reduces a population so it contains less genetic variability than the once-large population. A large population may become fragmented when a man-made dam creates a large lake where once an easily forded river provided no obstruction of movement. Likewise, a new highway can isolate species on either side. The net result is a small fragment that becomes reproductively separated from the main group. The fragmented group did not become isolated because of natural selection, so it may contain a fragment or all of the genetic variation of the larger population.

The *pioneer effect* occurs whenever a small group breaks away from the larger population to colonize a new territory. Like the fragmentation effect, the pioneers, which may consist of only a single seed or mating pair, remain an extinction threat because they do not have the genetic diversity of the main body and therefore are less likely to produce offspring capable of surviving changes in the environment. Even though a pioneer population reproduces successfully, the gene variation has not increased. So the danger involved with survival in a changing environment still exists!

Bionote

Even when small populations recover, their genetic variability is still so low that they remain in danger of extinction from a single catastrophic event.

Nonrandom Mating

The Hardy-Weinberg equation assumes that all males have an equal chance to fertilize all females. However, in nature, this seldom is true because of a number of factors such as geographical proximity, as is the case in rooted plants. In fact, the ultimate nonrandom mating is the act of self-fertilization that is common in some plants. In other cases, as the reproductive season approaches, the number of desirable mates is limited by their presence (or absence) as well as by their competitive premating rituals. Finally, botanists and zoologists practice nonrandom mating as they attempt to breed more and better organisms for economic benefit.

Bionote

Zoos spend a great deal of time, money, and energy in an effort to increase their genetic diversity by locating new breeding organisms for mating, usually from other zoos.

Speciation

According to the theory of natural selection, *speciation* is the creation of new species by genetic modifications of previously existing species, so the resulting organisms can no longer successfully mate and produce fertile offspring. Consequently, the most modern definition of species includes a retrieval of the genetic understanding from ancestral parents into a *biological species concept*, which states that a species is a population that can interbreed in nature and produce fertile offspring. New species have three principle mechanisms describing their formation, each of which involves reproductive isolation:

- Allotropic speciation
- Prezygotic isolation
- Postzygotic isolation

Allotropic Speciation

Suppose a volcano erupts and the overflow of lava blocks a stream to create a large lake. The resulting geographic separation may isolate organisms on either side of the lake as well as those above and below the dam. A physical separation that prohibits the gene migration between populations creates the opportunity for *allotropic speciation* for that subpopulation. When this happens, natural selection, mutation, and genetic drift act to genetically diversify the two populations so they are no longer capable of mating and producing fertile offspring. Geographic isolation presents the opportunity for the formation of a new species but cannot create a new species. True speciation only occurs when reproductive barriers prevent productive interbreeding. Two major types of *reproductive barriers* prevent a species from interbreeding even if they are in the same geographic area: *prezygotic* and *postzygotic reproductive isolation.*

Prezygotic Reproductive Isolation

There are five main types of prezygotic reproductive barriers that prevent intraspecies fertilization:

- **Territorial.** When two species are found in the same general area but not in the same habitat, they may not have an opportunity to mate. For instance, many species of plants live in the Amazon Basin rainforest; however, they are often stratified in their territories by competition and their need for light. Plants that require strong light grow only in the tree canopy, whereas low-light plants inhabit the shaded, forest floor, so the opportunity to interbreed never exists because of territorial separation.
- **Seasonal.** Species often have different mating seasons, which may prohibit mating opportunities. For instance, the wood frog breeds in late winter, whereas the common leopard frog breeds in early spring, and the common bullfrog breeds in early summer in most territories. Although sexually compatible in captivity, they do not interbreed in their natural setting because their reproductive seasons do not overlap!
- **Behavioral.** The "biological clock" in the natural state usually adjusts the breeding cycle to normal environmental pressures such as food production, availability of suitable nesting sites, mating opportunities, and predator activity. Different species often have different seasonal requirements that do not overlap.
- **Structural.** Species-specific behavior characteristics activate certain species without affecting the surrounding species. Different species often have various means by which the male and female of the species communicate their sexual readiness. The signals may be behavioral, such as the dance of the prairie chicken, or chemical, such as the scent or *pheromone* of your unspayed female dog. Even the vivid displays, such as the deep coloration change in tropical fish or the flagrant feather display of male peacocks, serve as specific readiness signals to potential mates. However, the mating ritual of one species may have no effect on a neighboring species even though their territories overlap. The inability of one species to recognize the mating signals of another species further increases their incompatibility and isolation. For instance, the ring-neck pheasant and the prairie chicken are similar birds that occupy similar habitats in the

Midwest. However, the pheasant is not excited by the premating dance of the prairie chicken and vice versa.

Structural limitations are prominent in certain flowering plants such that the actual plant structure, usually the flower, is designed in such a way that only a specific native pollinator has access to the pollen. A good example is the long, tubular shape of certain flowers that favor pollinators with long beaks. This mutualistic relationship ensures that the pollinators have diminished competition for that flower and the flower is assured that the specific pollinator will carry the pollen to like flowers to increase the likelihood of fertilization. This efficiency prevents wasted pollen and effort and prevents interspecies reproduction, which further isolates the plants. On a more visual level, a large dog may be reproductively attracted to a much smaller dog but find the structural orientation creates a barrier.

- **Genetic.** Even if the previous limitations are met, the genetics of a successful copulation do not always mean a successful fertilization. Even though sexual reproduction is designed to work the first time and every time thereafter, that rule applies to intraspecies mating. Successful interspecies copulation may produce gametes that are also incompatible and therefore do not unite to create a zygote, such as a hypothetical mating between a dog and a cat.

Postzygotic Reproductive Isolation

Three main barriers act on hybrid zygotes after interspecies fertilization:

- **Mortality.** Even if a hybrid zygote is formed, sometimes the inherited genes are incompatible, which then act to create weak or unhealthy offspring that do not survive to reproduce or are considered undesirable and not allowed to reproduce. In both cases, the genome for that organism fails to convey to the next generation and is lost. This represents a mortality barrier.
- **Infertility.** In a similar scenario, the resulting hybrid may reach mature adult status, capable of sexual reproduction, but is sterile. This represents an infertility barrier. Again, the inability to pass on the genes is a terminal genetic destination as exemplified by mules, the sterile product of a male donkey and a female horse. Because of their inability to breed and produce fertile offspring, horses and donkeys remain two separate species. A zebroid is the sterile offspring of a cross between a zebra and a horse. The zebra and the horse are in the same genus but not the same species.

Bionote

A whinny is the less-common, sterile product of a mating between a male horse and a female donkey.

- **Longevity.** Finally, in some cases, first-generation hybrid offspring are viable and do successfully interbreed and produce a second-generation offspring. Unfortunately, the second-generation offspring are inviable, either reproductively infertile or too unhealthy to reproduce. Regardless of the reason, in all cases, the longevity barriers promote genetic isolation and inhibit gene migration.

Sympatric Speciation

Sympatric speciation is the opposite of allopatric speciation because organisms, predominately plants, often create new species without the requisite geographic isolation. Plant-seed dispersal mechanisms often prohibit reproductive isolation, leaving sympatric speciation as the only major evolutionary cause agent for plants. Typically, a mutation occurs that prevents the offspring from successfully mating with a parent, but still allows viable reproduction with other individuals who inherited the same mutation. The most common mechanism is the chromosomal mutation that occurs because of a meiotic failure during gamete formation, when the chromosomes divide mitotically instead. When this happens, the duplicated chromosomes do not segregate and migrate into separate sex cells. Instead, they remain duplicated in the same sex cell, creating an overload of genes in certain gametes, which then become diploid, and deficient in others. It is possible then for the diploid gametes to unite with other diploid gametes to produce a *polyploid* individual, which contains more than the normal diploid complement of chromosomes. In plants, this occurs most frequently because of self-fertilization. The polyploid offspring can no longer successfully interbreed with the parent or any other similar-species organism that did not inherit the extra set of genes. *Sympatric speciation* is the reproductive isolation created by genetic abnormalities not as a result of geographic isolation. Although not widespread among animals, sympatric speciation has been significant in plant variation. Hugo de Vries, a Dutch botanist, is credited with identifying polyploidy as an agent of sympatric speciation. Through self-pollination, he created a large flowering polyploid evening primrose with 28 chromosomes instead of the normal diploid number of 14 chromosomes.

Speciation Rate

The speed by which new species are created depends upon the genetic makeup of the species, their ability to adapt to environmental changes, and the speed and severity of the environmental changes. In earlier times, it was thought that speciation occurred slowly over long periods of time. This *gradualistic theory* has recently given way to the *punctuated equilibrium model* that defines speciation as occurring in jumps or sudden shifts of speciation interspersed within long periods of inactivity. Emerging evidence from fossils lends support to the punctuated equilibrium model. This discussion is continued in the next section.

Macroevolution

Whereas microevolution explains diversification on an individual level over relatively short periods of time, *macroevolution* defines changes in large populations that often entail catastrophic environmental changes.

Geological Evidence

The fossil record establishes the ancestral lineage of both plants and animals and identifies periods of punctuated equilibrium in both. Rock strata can be used to date fossils because the organisms from which the fossils were derived died and were eventually buried in the material from which the rock was made. This allows a *relative dating* of the fossils by assigning their age in comparison with other rock strata. Except in the case of major Earth events, such as mountain building and erosion, the youngest rock strata and the fossils they contain are closest to the earth's surface and become older the deeper they are found in the crust. Also, rock strata in neighboring areas can be reconciled with each other if they are composed of similar rock or mineral type.

The evolutionary history for an area is arrayed in rock layers that jigsaw together to trace the *macroevolution* or major events of the history of life on Earth. As paleontologists discover fossils in a rock layer, they can make assumptions based on present-day life-forms about the environmental conditions that existed at that time. For instance, the discovery of a fern fossil would indicate a warm or temperate climate, adequate precipitation, and perhaps shade, all of which are conditions that support the growth of modern-day ferns. It may also provide clues that link with other fossil evidence to shed light on the animal life present at that time. For instance, an environment that supports ferns would also likely support herbivores like snails or an assortment of grazing animals that might feed on ferns.

Radiometric Data Analysis

Additional fossil evidence is collected using *radiometric* data analysis, which is a more approximate dating of once-living organisms by comparing the ratio of radioactive isotopes in their remains to that found in the atmosphere. Radiometric dating compares the ratio of the normal carbon-12 atom to the unstable, radioactive carbon-14 isotope. While alive, the ratio of carbon-12 to carbon-14 atoms in any living organism is nearly equal to their ratio in the atmosphere. Upon death, the carbon-14 is no longer added to the organism, and the existing amount begins to decay at a constant and known rate to a more stable isotope like nitrogen-14. The decay rate for all radioactive isotopes is called their *half-life*, which means that one-half of their mass will be converted, or decayed, into the more stable form in a known amount of time. So if you can determine the ratio in the sample and compare it to the atmosphere and then multiply by the known decay time, or half-life, you can establish the age. The half-life for carbon-14 is 5,600 years, meaning that one half of the beginning amount will decay to nitrogen-14 in 5,600 years. Radiometric analysis gives an actual age of the specimen and, when combined with rock strata data, can yield a more accurate time and location placement.

Bionote

Carbon-14 analysis is effective up to 50,000 years, after which other radioactive isotopes, such as potassium-40 and uranium-238, are used because they have a much longer half-life.

Geologic Time Scale

Macroevolution is often displayed on a *geologic time scale*, which highlights major evolutionary events in a comparative time scale. The smallest units of time on the geologic time scale are called *epochs* and measure in the millions of years, such as the Pleistocene epoch approximately two million years ago that included the Ice Age and the appearance of the first human fossils. Chronologically, epochs are clumped together into larger units called *periods*, such as Quaternary, which are combined to make *eras*, such as Cenozoic, which are the largest unit of macroevolution measure. The following table includes examples of macroevolution, from the oldest to the most recent.

Geologic Time Scale Example

Era	Period	Epoch	Event
Precambrian	—	—	Oldest prokaryote fossils.
Paleozoic	Ordovician	—	Origin of plants.
	Silurian	—	Land colonization by plants and arthropods.
Mesozoic	Jurassic	—	Dinosaurs roam.
	Cretaceous	—	Dinosaurs are extinct; flowering plants emerge.
Cenozoic	Tertiary	Paleocene	Radiation of mammals and birds.
	Quatenary	Pleistocene	Humans appear.

Phylogeny

Combined radiometric and rock-strata data analysis demonstrates evolutionary pathways for many plant and animal species. One of the most studied is the evolution of the modern horse, whose *phylogenetic tree* begins with a small, doglike creature and branches many times before reaching today's horse.

Bioterms

A **phylogenetic tree** is like a family tree or pedigree. It shows the ancestral relationships and genealogy for an organism.

Backmapping the phylogenetic tree to establish evolutionary links between fossils establishes the science of *systematics*—the organized scheme of classifying all living things into their phylogenetic tree.

The Least You Need to Know

- Fossil evidence supports abrupt levels of speciation, which are dated using geologic and physical methods.
- Pressures from the living and nonliving environment create an uneven selective pressure on reproductive success for all species.
- The Hardy-Weinberg equilibrium model mathematically integrates and explains allele frequencies in a population.
- Mutations are the physical rearrangements of the DNA creating new alleles, which is the ultimate source of genetic diversity.
- Multiple biotic and abiotic factors affect gene frequency and genetic diversity.
- There are both geographic and structural methods of speciation, which is the creation of new species. The rate of speciation is dependant on the DNA complement and the amount of environmental pressure.

Chapter 9

Systematics, Taxonomy, and Classification

In This Chapter

- Establishing the classification of animals
- Classifying unknown organisms
- Modern techniques for relating organisms
- How alternative forms of classification are beneficial
- Why a three-kingdom model is better than a six-kingdom model

Can you imagine describing a human as an "often hairy on top, bald on the bottom of walking surfaces, bipedal, eyes forward with binocular vision, opposable thumbs, no wings, no feathers, capable of rational thought, and the capacity to love"? In earlier times, such long, descriptive definitions were employed to separate plants and animals for identification. Can you then imagine the difficulty of combining these descriptions into a coherent grouping of like organisms for further study? Well neither could Carolus Linnaeus. He brought order, in fact a binomial order, to the classification scheme for identifying organisms. It's simple, two words: *genus* and *species*. You can create some of your own to describe your friends, such as *Homo beautifulus* or *Homo ignoramus;* you choose the names depending on the type of people you hang around with.

After organisms were cleverly classified, relations between and among them became more obvious. Analysis techniques used to establish kinship are presented; several are foolproof. However, that is where the understanding ends. Presently, there is a great amount of scientific discussion regarding whether the world would be better off by classifying organisms into a three-, five-, or six-kingdom model. No one has favored a four-kingdom model. All sides are presented here without editorial slant. You choose.

Linnaeus

Aristotle is credited with creating the first classification system more than 2,000 years ago. He classified living things as plant or animal according to their appearance. This system was expanded later by the Romans to be more specific down to the individual organism types, such as a cow or elm tree. As more organisms were identified and classified, more words were needed to separate similar organisms, such as two types of dog. So new organisms were classified by using lengthy descriptors, which eventually became cumbersome to use as more organisms created the need for more words. Carolus Linnaeus, a Swedish naturalist, began the struggle to classify all living things by proposing a *binomial*, or two-name system. In this model, the genus is the first name, and the species is the second name, with the first letter of the genus name always capitalized. Therefore, each organism is sorted by a genus, species classification, such as *Homo sapien* for man. Linnaeus also proposed to expand the genus-species nomenclature to include larger units of likeness, for recognizing extended degrees of kinship. Today, the categories are still based on Linnaeus's work. For example, the *taxonomy* for humans is the following:

Bioterms

Taxonomy is the classification of organisms, both plant and animal, based on their structural characteristics and evolutionary history.

Domain—Eukaryote

Kingdom—Animal

Phylum—Chordate

Subphyla—Vertebrate

Class—Mammal

Order—Primate

Family—Hominidae

Genus—*Homo*

Species—*Sapien*

Moving from domain to species, the organisms are more closely related, such that organisms in the same species have a greater degree of kinship than organisms that are similar only at their family level. For instance, both cats and humans are in the class *Mammalia*, but their pathways separate at that point because cats are in the order *Carnivora*, whereas humans are *Primates*. However, a grizzly bear, a black bear, and a polar bear are more closely related because they are in the same family: *Ursidae*. Further, grizzly bears and black bears are more closely related to each other than to the polar bear because the grizzly and black bear are in the same genus: *Ursus*, but the polar bear is not. Systematics operates to identify relative kinship and evolutionary intersects among species.

Examine the following classification chart. Which pair of organisms are most closely related?

	1	2	3
Kingdom	Animal	Animal	Animal
Phylum	Chordate	Chordate	Chordate
Class	Mammal	Mammal	Mammal
Order	Primate	Carnivore	Carnivore
Family	Hominidae	Felidae	Felidae
Genus	*Homo*	*Panthera*	*Felis*
Species	*Sapiens*	*Leo*	*Domesticus*

Organisms 2 and 3 are most closely related because they have the same family name. Organism 2 is a lion, organism 3 is a common housecat, and organism 1 is a human.

Binomial Classification

The binomial classification system proposed by Linnaeus allowed him and others to group organisms together based on common structures, functions, and resulting behaviors, which led to the science of taxonomy, or classification.

Often biologists use a *taxonomic key*, also known as a *dichotomous key*, to identify unknown organisms by their physical characteristics. Taxonomic keys work on a base-two premise: The organism either has the characteristic or does not. The resulting answer then directs the biologist to the next set of questions until all the characteristics have been accounted for and the organism is identified. The following *Binomial classification* illustration is a fictitious key that demonstrates the process.

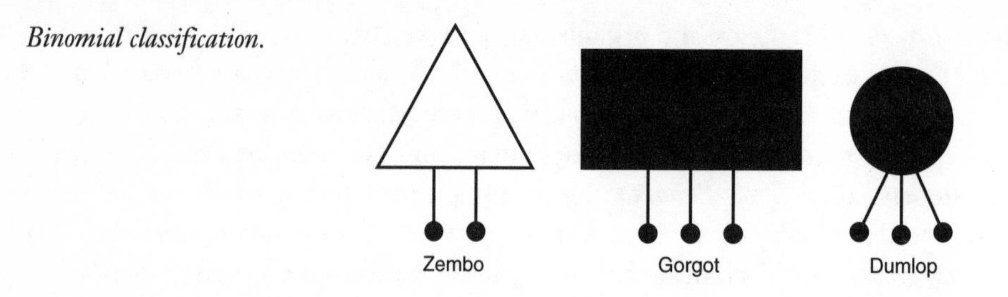

Binomial classification.

Try using the following taxonomic key to identify which organism is a "dumlop."

	Characteristic	Organism
1a.	It has two legs.	Zembo
1b.	It has more than two legs.	Go to 2
2a.	It is shaded.	Go to 3
2b.	It is not shaded.	Go to 4
3a.	It has a round head.	Dumlop
3b.	It does not have a round head.	Gorgot

Backtracking from the dichotomous key, a dumlop is an organism that has more than two legs, is shaded, with a round head. Most dichotomous keys are quite lengthy, to account for all the features that a set of organisms may possess. The broadest keys begin at the kingdom level and proceed to the more specific genus and species levels.

Systematics Analysis Techniques

Backmapping the phylogenetic tree for an organism's evolutionary pathway is the science of systematics. Systematics establishes degrees of kinship, diversity, evolutionary linkages, as well as taxonomic classification for all species.

A common method of establishing kinship is the identification of *homologous structures* shared by two different species.

Bioterms

A **homologous structure** is an anatomical term referring to commonalities in structural characteristics that evolved from the same feature in an ancestor.

For instance, a human arm, an alligator foreleg, and a penguin's flipper are homologous structures not because they look alike or function similarly, but because they both share a common bone structure indicating they all evolved from the same bone feature. The more homologous structures in common, the greater the degree of kinship.

Analogous structures are often confused with homologous structures because they appear to be similar in structure and function and resulting behavior. Convergent evolution occurs when dissimilar species living in the same territory and under the same environmental pressures develop analogous or similar structures that function in response to natural selection. As an example, both sharks and whales share common body structures that allow them to swim in the water. However, these structures are analogous, not homologous, because they trace their origin to different features: Whales are mammals, and sharks are fish! An embryonic analysis is often used to confirm homologous or analogous structures. Analogous structures do not indicate a degree of kinship.

Molecular Analysis

Anatomical analyses, such a homologous structures, have been the foundation of systematics until the onset of more sophisticated molecular analysis. Because this emerging technology has allowed for the identification and sequencing of proteins and nucleic acids, each organism can be individualized.

Nucleic Acid Analysis

The comparison of DNA nucleotide sequences between two separate specimens is the most accurate method of establishing kinship. Other methods include an analysis of rRNA nucleotide sequences and the extent of hydrogen-bonding similarities between DNA strands.

Amino Acid Analysis

Similarly, a comparison of the sequencing of amino acids in a polypeptide provides the most accurate analysis of protein structures. This analysis of protein backmaps to the DNA that coded for the amino acid sequence. The closer the match between amino acid segments, the more similar the DNA, the closer the phylogenetic relationship.

Alternative Methods of Classification

The Linnaean binomial system of classifying animals brought organization from chaos; but recently, with the application of modern technology, new methods have surfaced that yield additional information. Methods of establishing ancestral kinship are helpful in establishing new taxonomic procedures that often relate species in new ways. Although no one method is without drawback, each offers unique insights and information in reference to the organisms in question.

Cladistic Analysis

Cladistic analysis is probably the most widely used alternative method. Cladistic analysis is a means to classify organisms to match their evolutionary history. Common phylogenetic features are used to establish relatedness between organisms with the help of sophisticated computer programs that quickly sort organisms according to shared evolutionary structures.

Cladistic analysis sorts homologous structures into either a *primitive character* or a *derived character.* Primitive characters establish the broad classification that generates the basic grouping of organisms. For instance, a cladistic primitive character for plants is the presence of chloroplasts. Those organisms that contain chloroplasts are clumped into the same large grouping.

Derived characters are also homologous structures, but they represent features that have been modified for specific functions. Derived characters are more unique than primitive characters and tend to sort organisms by their presence or absence in the organism. The presence of a derived character or set of derived characters establishes a greater degree of relatedness. The more derived characteristics organisms share, the greater their degree of kinship. For instance, a derived characteristic in plants is the presence of vascular tissue. Advanced plants contain vascular bundles, but simple aquatic plants do not. This relatively simple anatomical feature demonstrates the vast difference between vascular and nonvascular plants (see Chapters 12 and 13). Review the example that follows to distinguish primitive and derived characters in mammals.

Mammalian primitive characters:

- Mammary glands and hair (for example, humans, dogs, cows, whales)

Derived characters:

- Appendages modified for aquatic movement (for example, whales)
- Appendages with an opposable thumb (for example, humans)
- Appendages designed for running (for example, dogs)
- Appendages designed for grazing on uneven ground and carrying heavy body weight (for example, cows)

After the primitive and derived characters are known, a cladogram can be constructed to show evolutionary linkages between groups of animals. Examine the illustration *Simple cladogram.*

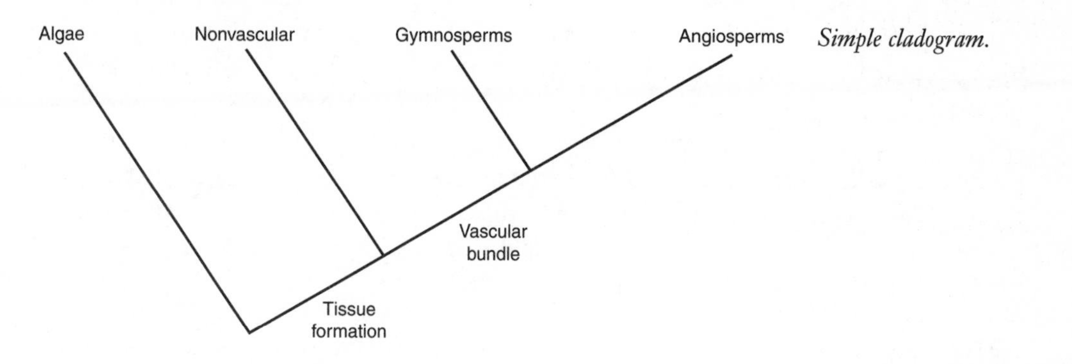

Simple cladogram.

This cladogram shows the evolutionary relatedness of major plant types by using simple derived characters on the right side in ascending order. Organisms located next to each other horizontally across the top are more related than those not in close proximity. For more specific classifications, such as the kinship between an oak tree and an elm tree, the cladogram would need more specific derived characters.

The cladistic model is somewhat similar to previous models of organization, except for several notable differences. The one most widely reported is the cladistic location of birds on a cladogram in relation to the reptiles. A cladogram relates birds closer to crocodiles and dinosaurs than to snakes or lizards. Interestingly, it is now known that reptiles did not evolve from a common reptilian ancestor, but are more likely descended from several different ancestors, making the reptile classification a conglomeration of animals with similar characteristics but dissimilar backgrounds! The cladogram correctly shows the hereditary linkage between birds and certain reptiles. It clearly is different from classical taxonomic classifications, which place all reptiles in one category (Reptiles) and all birds in another (Aves).

Cladistic analysis is extremely objective: The organism either has the feature or not. This strength is also a weakness. Opponents charge that the technique does not account for the amount or degree by which a feature is present or is utilized. In their opinion, this omission ignores too much relevant data and fails to make an accurate assessment of the difference between groups. For example, the fact that penguins have wings but do not use them to fly would create a cladistic analysis problem.

Evolutionary Systematics

A contrast to the no-bias approach of the cladistic analysis approach, the evolutionary systematic method deliberately builds in observer judgment. These taxonomists place heavier emphasis on the observed use or nonuse of a structure as well as the way it is used. Judgments are based on direct observation regarding the degree of evolutionary importance a particular feature is to that organism. For instance, in the previous bird and reptile analogy, the evolutionary systematic approach would lend more importance to the presence or absence of feathers than the derived homologous characteristics. So birds would be classified separate from reptiles. Most taxonomists agree that in the absence of data, the cladistic model is superior; with adequate data, the evolutionary systematic model has advantages.

Phenetics

Phenetics classifications are somewhat similar to evolutionary systematics in that both include all available data regarding the study of organisms and both are antagonistic to the cladistic analysis model. Phenetic classification does not attempt to establish evolutionary linkages but simply the clumping together of organism based on "overall" degrees of similarity. Phenetic classification requires access to the most data. Its overall effectiveness is diminished when the data are incomplete.

Modern Classification

Over the years, many models for classifying organisms have been touted as the next best one. Linnaeus's scheme of classifying everything into two kingdoms was the first real attempt, and it lasted for more than 200 years! However, because living organisms are so diverse, Linnaeus's classification system is considered inadequate today as a means of accurately representing the degree of relationship between organisms. More advanced thinking created first a five-, then a six-, then a three-kingdom classification system to include all the recent discoveries related to kinship. It is likely that as more discoveries are made and more data processed, the three-kingdom model may be in jeopardy.

Two-Kingdom Model: *Plantae, Animalia*

Linnaeus's model classified everything as either a plant (kingdom *Plantae*) or animal (kingdom *Animalia*), which was sufficient for that era. However, even Linnaeus's followers struggled with the classification of fungi. Most placed them in the plant kingdom because they appeared to have roots and certainly did not belong in the animal kingdom! As additional species were discovered, the two-kingdom classification gave way to other models.

Five-Kingdom Model

American ecologist Robert Whittaker created a five-kingdom model that accounted for prokaryote and eukaryote distinctions. In this schemata, all prokaryotes were placed in the kingdom *Monera.* The remaining eukaryotes were separated by differences mostly in structure. Plants and animals were easily separated into their own kingdoms. It was decided that because fungi were neither plant nor animal but subsisted by decaying or decomposing once-living matter, they also become their own kingdom. Everything else was clumped into the kingdom *Protista.* The protists included all eukaryotes that did not clearly fit into the plant, animals, or fungi kingdoms. With the advent of sophisticated collecting and microscopy techniques, the question arose as to where the "new" types of bacteria belong.

Six-Kingdom Model

Linnaeus created his model during the eighteenth century, which proved useful for scientific knowledge at that time. However, recent discoveries of archaebacteria in deep-ocean thermal vents, hot springs in Yellowstone, and brine marine environments justified another look at the five-kingdom model. Analysis of archaebacteria indicates that they are more similar to eukaryotes than their bacterial prokaryotic cousins (see Chapter 11). The six-kingdom model contains the same original four kingdoms, but splits the *Moneran* kingdom into kingdom *Archaea* bacteria to include the newly discovered bacteria type with internal membranes, and kingdom *Eubacteria,* which contains the more common forms of bacteria. Interestingly, cladistic analysis indicates that organisms in the kingdom *Protista* are actually more different from each other than originally suspected. In fact, most mammals are more closely related to certain archaebacteria than some protists are related to each other!

Three-Kingdom Model

The five- and six-kingdom models are based on the scheme of two fundamentally different groups of organisms: prokaryote and eukaryote. They divided and sorted all living things into the various kingdoms based on similarities of evolutionary history. The three-kingdom model makes a fundamental change. Instead of classifying organisms into prokaryote or eukaryote, it sorts them into three categories by splitting the prokaryotes into two kingdoms similar to the six-kingdom model: *Archae* bacteria and

Eubacteria. All the eukaryotes, which is virtually everything else, are clumped into the *Eukarya* kingdom. Modern chemical and cellular evidence currently supports the three-kingdom model. It is believed that bacteria were among the first type of life, and the eubacteria separated leaving the archaebacteria and eukarya to continue as one in evolutionary struggles. The fact that modern archaebacteria contain many eukaryotic features indicates their separation occurred after the eubacteria and thereby establishes greater kinship between archaebacteria and eukarya. Archaebacteria are established as a midpoint between eubacteria and eukaraya. Most modern taxonomists favor the three-kingdom model.

Perhaps as more species are discovered, additional anomalies will occur that force the reorganization of the model. We may end up with a four-kingdom classification model!

The Least You Need to Know

- Linnaeus developed a schema for organizing the diversity of plants and animals.
- A dichotomous key can be used to sort organisms into like groups.
- Genetic analysis establishes an accurate basis for establishing the degree of kinship between organisms.
- Homologous structures indicate a common ancestral structure, but analogous structures do not indicate a relationship.
- Cladistic analysis is one of several alternative methods of classification.
- Currently, most taxonomists are using a three-kingdom model, although six-kingdom and five-kingdom models are available.

Chapter 10

Origin of Life

In This Chapter

- Evidence that finally disproved spontaneous generation
- Influences on the origin of life theories
- Geologic evidence to support the theories of life genesis
- Life created in a laboratory

You may have already know that somebody somewhere has mixed a few chemicals together, ignited it with a spark, and created life, or at least amino acids. Science fiction or science fact? You decide after you read what Dr. Miller did in his lab. Is it possible to create life from nonlife? Redi, Spallanzani, and Pasteur all created experiments that disproved this hypothesis. So how did life start? Several easy-to-remember ideas are presented in this chapter for you to enjoy.

On top of that, have you ever noticed how the east side of South America looks like it could fit into the west side of Africa like a jigsaw puzzle? If so, you are not alone. If not, go look at a map more closely. Wegener thought that perhaps they were once joined and somehow separated. The learned people of Wegener's time thought *he* was separated … from his good sense. We now know that he was right. You and I are riding on the top of large chunks of real estate that are slowly moving. The movement of these vast continental plates has been the speculated source of massive extinctions followed by periods of immense speciation. Curious? Read on.

Spontaneous Generation

It was once believed that life could come from nonliving things, such as mice from corn, flies from bovine manure, maggots from rotting meat, and fish from the mud of previously dry lakes. Spontaneous generation is the incorrect hypothesis that nonliving things are capable of producing life. Several experiments have been conducted to disprove spontaneous generation; a few of them are covered in the sections that follow.

Redi's Experiment and Needham's Rebuttal

In 1668, Francesco Redi, an Italian scientist, designed a scientific experiment to test the spontaneous creation of maggots by placing fresh meat in each of two different jars. One jar was left open; the other was covered with a cloth. Days later, the open jar contained maggots, whereas the covered jar contained no maggots. He did note that maggots were found on the exterior surface of the cloth that covered the jar. Redi successfully demonstrated that the maggots came from fly eggs and thereby helped to disprove spontaneous generation. Or so he thought.

In England, John Needham challenged Redi's findings by conducting an experiment in which he placed a broth, or "gravy," into a bottle, heated the bottle to kill anything inside, then sealed it. Days later, he reported the presence of life in the broth and announced that life had been created from nonlife. In actuality, he did not heat it long enough to kill all the microbes.

Spallanzani's Experiment

Lazzaro Spallanzani, also an Italian scientist, reviewed both Redi's and Needham's data and experimental design and concluded that perhaps Needham's heating of the bottle did not kill everything inside. He constructed his own experiment by placing broth in each of two separate bottles, boiling the broth in both bottles, then sealing one bottle and leaving the other open. Days later, the unsealed bottle was teeming with small living things that he could observe more clearly with the newly invented microscope. The sealed bottle showed no signs of life. This certainly excluded spontaneous generation as a viable theory. Except it was noted by scientists of the day that

Spallanzani had deprived the closed bottle of air, and it was thought that air was necessary for spontaneous generation. So although his experiment was successful, a strong rebuttal blunted his claims.

Pasteur's Experiment

Louis Pasteur, the notable French scientist, accepted the challenge to re-create the experiment and leave the system open to air. He subsequently designed several bottles with S-curved necks that were oriented downward so gravity would prevent access by airborne foreign materials. He placed a nutrient-enriched broth in one of the goose-neck bottles, boiled the broth inside the bottle, and observed no life in the jar for one year. He then broke off the top of the bottle, exposing it more directly to the air, and noted life-forms in the broth within days. He noted that as long as dust and other airborne particles were trapped in the S-shaped neck of the bottle, no life was created until this obstacle was removed. He reasoned that the contamination came from life-forms in the air. Pasteur finally convinced the learned world that even if exposed to air, life did not arise from nonlife.

Bioterms

Pasteurization originally was the process of heating foodstuffs to kill harmful microorganisms before human consumption; now ultraviolet light, steam, pressure, and other methods are available to purify foods—in the name of Pasteur.

Early Earth Environment

So if Pasteur is correct and life only comes from existing life, where and how did life begin? Many theories attempt to answer this question, including the popular creationist theory, which states that God created man in his own image, which may in fact be correct. However, this section illustrates the scientific evidence that leads to an evolutionary pathway. In the final analysis, both theories may turn out to be the same.

Based on many assumptions, the conditions on early Earth, some three to four billion years ago, are thought to be much different from what they are today. To begin with, the astronomical phenomenon called "the big bang" is defined by a theory proposing that the earth was one of the larger particles that coalesced after the initial universe explosion, or big bang, that spewed all the particles in the universe away from a central point and destined them to slowly revolve around that point.

Consequently, the earth was very hot, evaporating the liquid water into the atmosphere. However, as the earth cooled, gravity-trapped water vapor condensed, fell as rain, and did not boil away but remained impounded in pools that became lakes and oceans. It was also believed that tectonic activity caused many volcanic eruptions at that time. From present-day volcanoes, we know that when they erupt, they release carbon dioxide, nitrogen, and a host of nonoxygen gases. In addition, with no protecting atmosphere, the earth was constantly bombarded with meteorites and other space debris still in circulation from the big bang. From current astronomical research, we know that meteorites can carry ice and other compounds, including carbon-based compounds. Researchers believe, therefore, that early Earth's atmosphere consisted of water vapor, carbon dioxide, carbon monoxide, hydrogen, nitrogen, ammonia, and methane. Note that no oxygen was present in early Earth's atmosphere!

Bionote

Iron-containing rocks reportedly recovered from period strata contain no rust, further indicating the absence of oxygen.

Meteorologists suspect that lightning, torrential rains, and ultraviolet radiation combined with the intense volcanic activity and constant meteorite bombardment to make early Earth an interesting but inhospitable environment.

Miller-Urey Synthesis

Two American scientists, Stanley Miller and Harold Urey, designed an experiment to simulate conditions on early Earth and observe for the formation of life. They combined methane, water, ammonia, and hydrogen into a container in the approximate concentrations theorized to have existed on early Earth. To simulate lightning, they added an electrical spark. Days later, they examined the "soup" that formed and discovered the presence of several simple amino acids! Although this experimental design probably did not accurately represent early Earth's percent of gaseous combinations, further work by Dr. Miller and others, using different combinations, all produced organic compounds. As recently as 1995, Miller produced uracil and cytosine, two of the nitrogen bases found in both DNA and RNA. However, to this date, no living things have been made from nonliving things in the laboratory. Interestingly, continuing research on meteorites has identified, as recently as 1969, that they contain all five of the nitrogen bases. This presents the hypothesis that perhaps the ingredients necessary for life were brought from outer space!

Wegener: Plate Tectonics and Continental Drift

By looking at a modern-day map of the world, it is easy to see how the coastline of the west side of Africa appears to match the east coastline of South America.

As cartographic skills and knowledge of the continent's boundaries increased by nautical exploration, in 1912, German meteorologist Alfred Wegener proposed an Earth-moving hypothesis. He hypothesized that the existing landmasses are actually moving and probably all began as one large landmass. His theory of *continental drift* made the landmasses of Earth appear like giant floating islands sometimes moving away, sometimes crashing into each other by forces he could not describe. Although the Africa–South America anomaly was noted, his theory did not gain much support in his lifetime.

With recent advances in geology, we now know that all the surface features—land and water—are actually floating on the viscous mantle of the earth, which supports the movable crust and outer layer of Earth. The solid crust, or plate, that we inhabit is one of many irregularly shaped pieces of varying size that move in specified directions. The idea that these large continental plates are in constant motion created by geothermal heating, convection, and movement is called *plate tectonics.*

Bionote

Most of North America and about one-half of the adjacent Atlantic Ocean ride on the North American plate. The large Pacific plate rubs against the North American plate at the San Andreas fault in Southern California, creating frequent earthquakes.

Plate tectonics explains how large landmasses separate and also collide into each other. This constant Earth movement, often measured in centimeters per year, is responsible for earthquakes, volcanoes, sea-floor spreading, and continental drift.

Apparently, Wegener was correct; the early isolated land forms probably joined together to create a single landmass, or supercontinent called *Pangaea,* approximately 250 millions years ago at the end of the Paleozoic era. Note in the illustration *Pangaea* the proposed shape of the supercontinent.

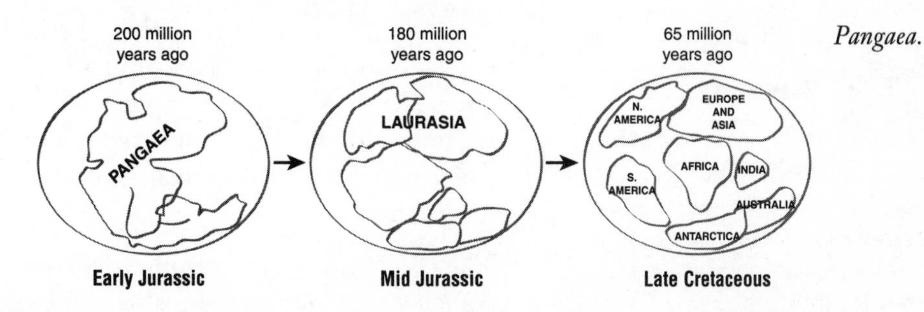

Pangaea.

Life that had evolved on the separate landmasses now had to compete with other life-forms from the other isolated landmasses as these landmasses congealed into one.

Competition for space, food, and shelter as well as increased predation created additional natural-selection pressures. Fossil records indicate mass extinctions and a major change in genetic diversity at this time.

A second cataclysmic event also affecting biological diversity occurred about 200 million years ago during the Mesozoic era. At that time, Pangaea began to separate, and the isolated land forms again became their own unique isolated evolutionary laboratory. The separating landmasses became reproductively isolated from one another.

Extinction and Genetic Diversity

Extinction appears to be a natural phenomenon and, like natural selection, favors the reproduction of certain species at the expense of less-fit species. Extinction is the loss of all members of a given species and their genetic complement, never to be recovered. Fossil evidence indicates that following a mass extinction such as the Permian extinction, when Pangaea was formed; and again at the end of the Cretaceous period when dinosaurs ruled the world, a period of growth and genetic variation followed. Apparently, the extinctions opened the fringe territories for colonization by the remaining species. Mammals are the classic study on this point because they were known to exist 50 to 100 million years in territories inhabited by dinosaurs before the extinction of the dinosaurs. Following the demise of the dinosaurs, mammalian fossils indicate a considerable amount of speciation and growth in overall numbers, both probably associated with the acquisition of new territory and the loss of dinosaurs as competitors and predators.

Adaptive Radiation

The rapid genetic diversity following an extinction, landmass split, or other cataclysmic event may be due to *adaptive radiation*, also known as *divergent evolution*.

Bioterms

Adaptive radiation is the process by which genetic diversity is increased in descendants of a common ancestor as they colonize and adapt to new territories.

It is called radiation because the genetically divergent descendants appear to radiate from a central point, much like the solar rays from the sun. During divergent evolution, descendants adopt a variety of characteristics that allow them to occupy similarly diverse niches.

The classic example of adaptive radiation is the study completed by Darwin as he observed 13 different finch species during his famous voyage of discovery to the Galapagos Islands. The islands themselves are well suited for adaptive radiation because they consist of numerous small islands in close proximity in the Pacific Ocean approximately 125 miles (200 kilometers) west of Ecuador, South America.

Since Darwin's time, an analysis of the finch speciation revealed a *founder population* arrived from the mainland and occupied an island. Specific island pressures probably caused that species to evolve into a new species different from the mainland species. As the finches overtook the island, competition increased, and pioneer species may have migrated to a different island. This created a new founder species that adapted to the new island pressures and modified to become a new species. Likewise, the remaining islands were colonized in succession. Because each island is slightly different, the finch adaptations were often unique to a specific island. In addition, finches could return to an inhabited island and compete with the existing species, or return and divide territory, shelter, and resources and peacefully coexist. The return to an inhabited island also probably sparked additional natural-selection pressures.

We are still not sure how life originated on Earth. It could be a heavenly masterpiece, an astronomical anomaly, or a series of mutations and adaptations. There is evidence that favors each theory. Regardless, patterns in similarity appear to link some organisms more closely than others.

The Least You Need to Know

- Spontaneous generation is an incorrect theory that states that life can start from nonliving substances.
- Louis Pasteur conducted the conclusive experiment that demonstrated that life could not arise from nonlife.
- The early earth environment was a violent location where catastrophic events such as volcanoes and meteor bombardments occurred frequently in an oxygen-free atmosphere.
- Dr. Miller has created amino acids in the laboratory using raw materials that were thought to be present on early Earth.
- Alfred Wegner's proposal that large, continent-sized landmasses move in specific directions became the basis for the modern theory of plate tectonics.
- Extinctions caused by geographic, cataclysmic events created expansive genetic diversity among the surviving animals through many processes, especially adaptive radiation.

Chapter 11

Origin of Prokaryotes and Eukaryotes

In This Chapter

- Early organism replication without DNA
- How the prokaryotes evolved into eukaryotes
- Functional diversification
- Classification of bacteria and archaea bacteria
- How protists resemble plants and animals
- Why slime molds change their shape

Where is RNA World? It sounds like a theme park. Actually it is the proposed first step in transitioning from nonliving to living. Several hypotheses describe the assimilation of chemicals into an order that responds to the environment in a living sort of way. If we assume those ideas are correct, the next step is the world of the living.

Talk about living … the archaebacteria and their one-of-a-kind lifestyle are a frontier area for science. Other lifestyles are presented in this chapter as well, as we look at the different types of small things that inhabit the world around us. Some of them change shapes on a regular basis, making them appear to be two different organisms! This is fascinating biology.

Origin of Prokaryotes

In 1862, Pasteur disproved the spontaneous-generation theory but left open a question: How did life begin? Miller's synthesis is a possible answer, or it may be the seeding of organic molecules by meteorites from outer space, or a God event that started life. It is generally held that the first organisms were formed around four billion years ago, with the earliest forms being simple molecular groupings that somehow gained the ability to metabolize and reproduce. It is also held that these simple molecular arrangements formed from existing inorganic substances—life from nonlife!

RNA World

RNA World describes the hypothetical time of the earliest life-forms when genes were simply strands of RNA. It is interesting to note that only nucleic acids have the ability to replicate and store genetic information, one of the fundamental characteristics of life. Referencing that certain prokaryotes and viruses do not contain DNA, but reproduce solely with RNA, it is also believed that the earliest life-forms were nucleic acids that simulated RNA in structure and function. Laboratory experiments confirmed that nucleotide monomers can spontaneously join to form gene-like structures composed of RNA segments. They can also create the complementary strand of RNA.

The strands of RNA were then available to serve as a template to bind amino acids together into polypeptides. Most long molecules, such as proteins, respond to the surface tension of water in an aquatic environment to spontaneously form circles and rings, called *microspheres*. These nonliving microspheres appear to function as a cell membrane. They have been shown to grow by adding monomers and divide as they become too large, and also demonstrate some selective permeability by allowing water-soluble substances to pass while prohibiting fat-soluble transport. Scientists hypothesize that these early rings may have surrounded RNA segments of genes to form a cooperative alliance. *Coacervates* are droplets of organic molecules that include amino acids and sugars. Both coacervates and microspheres spontaneously form into spheres under certain conditions. These conditions are thought to be similar to early Earth, such as a hot surface (a sunbaked or geologically heated rock, for instance).

The heat provides the energy for the dehydration synthesis, which joins basic units together to make more complex molecules. Note that genes were not required!

Chemiautotrophic Prokaryotes and the Heterotroph Hypothesis

The advancements in prokaryote complexity may have evolved from a more efficient acquisition of food. Based on studies of archaebacteria, scientists theorize that the earliest prokaryotes absorbed energy from extracellular reactions to power the formation of ATP within the cell. These *chemiautotrophic* cells probably used carbon dioxide as the carbon source and the energy of ATP to construct larger and more complex molecules. Another theory, called the *heterotroph hypothesis*, suggests that the aquatic environment was full of organic molecules, including ATP, which were then absorbed into the cell for cellular functions. The first heterotrophs could have survived easily on the supposed soup of organic molecules in their consumptive environment. However, at some point the heterotrophs would inevitably exhaust their food supply and the autotrophic system would replace the heterotrophs and become established as the dominant life-form. In both cases, the presence of enzymes is necessary, and their origin is not fully understood.

Today prokaryotes are found everywhere life exists on Earth and greatly outnumber all eukaryotes combined. Prokaryotes contribute as decomposers and recyclers to such an extent that without them, eukaryotes would die off. However, prokaryotes could survive without eukaryotes as they have already demonstrated for about two billion years!

Bionote

Prokaryotes are also assumed to be the life source that first performed photosynthesis, which then contaminated the atmosphere with oxygen, which selected against anaerobic metabolic pathways and caused one of the greatest mass extinctions.

Prokaryote Evolution: Bacteria and Archaea

Prokaryotes are mostly bacteria, and their advancements led to more complex living organisms. It has been suggested that the diverse nature of bacteria and archaebacteria resulted from this evolution. As bacteria modified structures to expand their territory and tolerance, they changed into newer species of bacteria with diverse structures and functions. Due to their uniqueness, bacteria are classified in their own kingdom!

Advancements in the structure and function of prokaryotes continued to the juncture where two separate types are now identifiable: bacteria and archaea.

Bacteria and Cyanobacteria

Bacteria are the most common and well studied because they are the easiest to find and have historically been the source of many human maladies, such as bubonic plague, tuberculosis, and cholera, and the source of much advancement such as cheese, recombinant DNA, and intestinal flora, which aids in digestion and nutrient production.

Bacteria appear to be simpler than archaea because they do not possess certain advanced structures typical in archaea, such as the complex RNA polymerase, the presence of interons, and branched carbon chains in lipid membranes, as well as some internal membranes. However, they do possess a cell membrane and have definite life functions. They exist alone or in colonies, in a variety of shapes, and some can endure unfavorable conditions by forming a protective *endospore* around the cell, which allows the cell to remain viable and dormant until favorable conditions arrive. Bacteria and archaea do possess whiplike flagella for movement.

Bionote

Even today, anabaena, a typical cyanobacteria, blooms in nutrient overloaded aquatic environments to produce a telltale blue-green color. Environmentalists use anabaena blooms as an indicator of environmental quality.

Cyanobacteria, also known as blue-green algae, are intriguing organisms because they contain photosynthetic capabilities and are thought to be responsible for changing the prehistoric environment to an oxygen atmosphere.

Microfossil cyanobacteria estimated to be 3.5 billion years old were discovered in Australia. Their hypothesized oxygen production likely also created the protective ozone layer.

Archaea

Archaea have structures such as tRNA nucleotide sequences and RNA polymerase that are more closely related to eukaryotes than bacteria. They have adapted complex protein, carbohydrate, and lipid molecules that allow them to live and reproduce in the harshest environments where nothing else will live. In fact, archaea are so different from bacteria that they are also classified in their own kingdom, separate from all other organisms! Many species are autotrophic and obtain energy through the chemosynthesis of carbon dioxide instead of the photosynthesis of carbon dioxide. Because of their extreme lifestyle, they do not have the history of scientific investigation that bacteria have generated, although they contain the solutions for expanding the genetic territory of other helpful microorganisms. For example, archaebacteria thrive in the hot springs in Yellowstone National Park where the water temperature is measured at 194°F (90°C).

Eukaryote Evolution

Fossil records indicate that eukaryotes evolved from prokaryotes somewhere between 1.5 to 2 billion years ago. Two proposed pathways describe the invasion of prokaryote cells by two smaller prokaryote cells. They subsequently became successfully included as part of a now much larger cell with additional structures and capable of additional functions.

- Endosymbiosis
- Membrane infolding

Endosymbiosis

Research conducted by Lynn Margulis at the University of Massachusetts supports the hypothesis that two separate mutually beneficial invasions of a prokaryote cell produced the modern-day mitochondria and chloroplast as eukaryotic organelles. In this model, ancestral mitochondria were small heterotrophs capable of using oxygen to perform cellular respiration and thereby create useful energy. They became part of a large cell either by direct invasion as an internal parasite or as an indigestible food source. Later, a second invasion brought ancestral chloroplasts, which are thought to be small, photosynthetic cyanobacteria. Modern-day supporting evidence for endosymbiosis shows that both the mitochondria and chloroplasts have their own genes, circular DNA and RNA, and reproduce by binary fission independent of the host's cell cycle. They therefore appear to be more similar to prokaryotes than eukaryotes.

Membrane Infolding

The invasions of the host prokaryote cell probably were successful because the host cell membrane infolded to surround both invading prokaryote cells and thereby help transport them into the cell. The membrane did not dissolve but remained intact, and thereby created a second membrane around the protomitochondria and protochloroplast. It is also known that in modern-day eukaryotes the inner membrane of both the mitochondria and chloroplast contain structures more similar to prokaryotes than eukaryotes, whereas the outer membrane retains eukaryote characteristics! It is also suggested that continued membrane infolding created the endomembrane system.

It can be said that possibly the first eukaryotic cell type was miraculously born from prokaryotic, symbiotic, multicell interactions!

Protists

Protists represent in excess of 100,000 species and are so varied in their structure and function that originally some were considered plants, others animals, others fungi, and some, a combination. As such, the kingdom Protista is often described as representing those organisms that are eukaryotes, not plants, not animals, and not fungi. Although most are unicellular, several, such as the giant kelp, are multicellular but lack specialized tissues. The protists may also represent the ancestors of modern-day plants, animals, and fungi.

Animal-Like Protists

There are four separate phyla of protists with animal characteristics. In early classification schemes, they were clumped together and called *protozoa* to separate them from the more plantlike protists:

- Sporozoa
- Sarcodina
- Ciliata
- Zoomastigina

Sporozoa

Sporozoa are among the best known protists because they are all parasites, including human parasites. They usually live in a host organism and reproduce by spores, which are dormant cells enclosed in a protective membrane. Whenever the spores land on an appropriate host, they are able to enter by various means and then grow to maturity as parasites. The parasitic cells have specialized organelles for penetrating host cell membranes. More exotic sporozoa have life cycles that involve two hosts.

Bionote

Malaria is still a problem in developing tropical countries where millions of people become infected annually.

The *plasmodium* is the parasite that causes malaria by entering human red blood cells and digesting their nutritional contents until the red blood cells become nonfunctional. The plasmodium then grows, reproduces, and infects neighboring red blood cells. Occasionally a female *anopheles* mosquito will withdraw some plasmodium-infected blood as part of her normal dietary requirement and then transfer the plasmodium to another unsuspecting victim.

Sarcodina

The phylum *Sarcodina* is best known for their bloblike structures, called *pseudopodia*, that provide a means of locomotion. Pseudopods are temporary membrane-bound cytoplasm projections that direct the motion of the sarcodines. This innate flexibility allows the sarcodines to assume virtually any shape. Amoebas are typical sarcodines that use pseudopods to locate, surround, and engulf food sources. Other interesting examples include the *foraminifera*, which are aquatic protists mostly known by the calcium carbonate shells they secrete, which sometimes accumulate in large deposits when they die, such as the famous White Cliffs of Dover, England. Because foraminifers only inhabit warm waters, whenever a geologist discovers a strata containing their fossilized shells, the climate for that aquatic environment at that time can be estimated fairly accurately.

Ciliata

The *ciliates* (*phylum Ciliophora*) exhibit several advancements not associated with the previous protists. They exist as free-living, nonparasitic, fresh- or saltwater, unicellular or colonial organisms. They also have developed short, hairlike structures, called *cilia* that move in rhythm for locomotion. Cilia are often described as functioning similar to oars for movement of a ship, which is accurate except cilia sometimes surround the entire organism. They allow directed movement toward a food source and away from inhospitable territories. Paramecia are a common example of a ciliate and exhibit another interesting phenomenon—they have two nuclei. A large macronucleus controls the everyday activities of the cell, and a smaller micronucleus (often more than one) functions during gamete exchange. Under normal conditions, paramecia reproduce asexually by binary fission (refer to Chapter 2); during periods of stress, however, they *conjugate*, meaning they exchange haploid micronuclei with another paramecium. Refer to the illustration *Paramecium conjugation* for a pictorial representation.

Because no offspring or fertilized eggs are produced, technically, sexual reproduction did not occur, but gametes were exchanged by mature adults, resulting in a new genetic complement for both paramecia! Paramecia also contain most organelles that more advanced life-forms utilize. For instance, in addition to the mitochondria and nucleus, they also use food vacuoles containing digestive enzymes, an anal pore for waste removal, and contractile or water vacuoles for water transportation.

Paramecium conjugation.

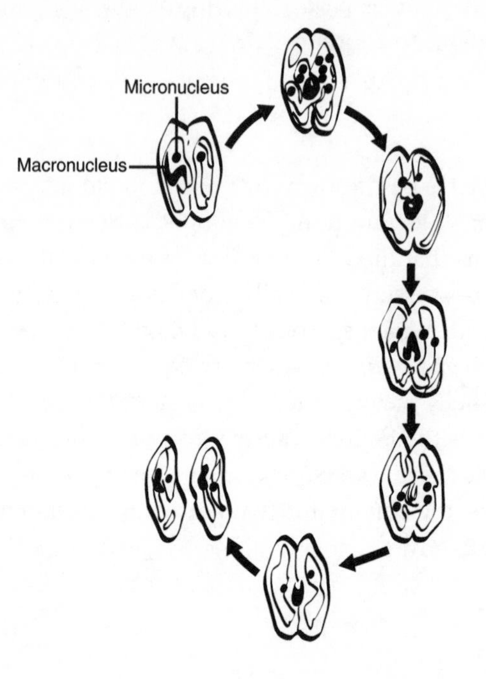

Zoomastigina

Zoomastigina, also known as flagellates, are known for their specialized *flagella*, which are whiplike structures that propel the flagellates through their aquatic environment. Typically flagellates have only one flagella, but may have up to four working in sync. Although most flagellates are harmless, simply surrounding and engulfing their food, others are human parasites. One of the most interesting parasites is the *trypanosome*, which causes African sleeping sickness. The symptoms are well known: fever, chills, and skin rash. Affected individuals become very weak, unconscious, and may fall into a fatal coma.

The trypanosome is transmitted by the tsetse fly and lives in the bloodstream, and continually change their surface molecular structure to gain invisibility to the host's immune system and remain undetected in attacks on the host. The disease attacks the nervous system of infected individuals.

Plantlike Protists

The three plantlike protists all contain chlorophyll, and as autotrophs utilize photosynthesis to make their own energy. They are generally multicellular and mobile, usually flagella assisted, inhabit wet or aquatic territories, and do not have true roots, stems, or leaves, but are considered to be a type of algae:

- Euglenophyta
- Chrysophyta
- Pyrrophyta

Euglenophyta

Euglenophytes are also known as green algae and are structurally similar to Zoomastigina because both utilize flagella and have common structures. However, the euglenophytes also contain chloroplasts and undergo photosynthesis. The euglena is a typical euglenophyte, even lending its name to the phylum. Euglenas contain an eyespot, which does not focus as an eye, but does differentiate light from dark and allows a euglena to move toward a light source for greater photosynthetic opportunities. They are also capable swimmers with two flagella, which is important because they can then inhabit diverse aquatic territories. Living as both a photosynthetic autotroph and, in the absence of light, as a nutrient absorbing heterotroph, the euglenas are quite versatile in their territory requirements. Their range is also not limited by reproduction, which they accommodate both sexually, whenever possible, and asexually the rest of the time. Refer to the illustration *Typical euglena.*

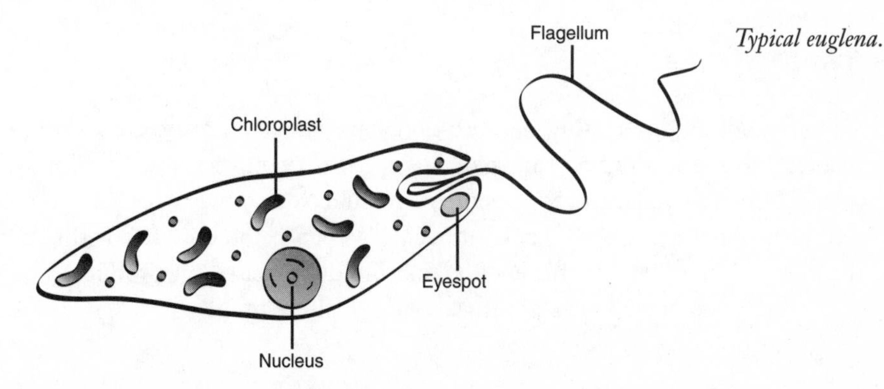

Typical euglena.

Volvox is an example of a colonial euglenophyte consisting of individual cells resembling a hollow ball, which is also capable of producing daughter colonies. Volvox cells and their cousin *chlamydomonas* contain features in common with more complex plants such as cell walls made of cellulose, starch as an energy-storing compound, and chloroplasts, suggesting these protists may have evolved into modern-day plants.

Chrysophta

Diatoms are the principle species found in the phylum *Chrysophyta*. Diatoms are unique because they store their food in a light, less-dense oil form that allows them to float on top of water to be closer to sunlight. They also substitute the carbohydrate pectin for cellulose in their cell wall, which is rich in the element silicon, the main ingredient in glass. Diatoms are an important freshwater and marine food source; when dead, their shells often accumulate in large strata known as diatomaceous earth, which is still used as an abrasive and for filtering purposes.

Pyrrophyta

Dinoflagellates are unicellular photosynthetic algae that are the principal member of the phylum *Pyrrophyta*. They exhibit three unusual properties. First, several species are luminescent and give off light when disturbed. Second, their DNA is not packed with histones, making them separate from all other eukaryotes. Third, dinoflagellates are subject to population explosions whenever conditions are favorable. This "bloom" is sometimes called a "red tide" because the algae are red in color and are so numerous that the water appears reddish. They are often so thick they block the absorption of oxygen by less-mobile aquatic life, such as flounder and crabs. Often these animals move to the more oxygen-rich waters near the shore, where they contribute to the human jubilee on the shore.

Slime Molds

Slime molds are interesting because they have a dual identity, each with separate characteristics. During part of their life cycle, they appear as small blobs of protoplasm that are similar in movement to an amoeba as they surround and engulf bacteria and other available organic nutrients. When harsh conditions threaten, they form into a completely different-looking moldlike mass that produces funguslike reproductive spores. There are three types of slime molds:

- Oomycota
- Myxomycota
- Acrasiomycota

Oomycota

Oomycota slime molds are all parasitic and are usually aquatic; however, some are also found in the soil. These molds are responsible for the white, threadlike growths seen on diseased tropical fish in home aquariums. They also caused the great Irish potato famine that led to the deaths of more than 400,000 people. The Irish potato famine was actually a series of crop failures that occurred in Ireland during the years of 1845, 1846, 1848, and 1851. The Irish economy was heavily based on the production and sale of potatoes. When the blight attacked the young plants, the tubers (potatoes) rotted so no crops were available to generate income. As a result, widespread starvation ensued with an estimated one million people emigrating to the United States or Great Britain.

Myxomycota

Acellular slime molds, *Myxomycota*, are unique in producing structures called *plasmodia*, which appear to be a mass of cells, but in reality are one large cell that contains many nuclei free to float unhindered by internal membrane barriers. The plasmodium slime mold remains in this blob stage feeding on bacteria, fungi, and other organic nutrients, typically found on the floor of a deciduous forest, because of its mobility. In its blob form, it also has greater surface area to transfer materials to and from the environment. However, when food or water is scarce or the environment becomes too dry, the plasmodium transfigures into the moldlike mass and produces the characteristic fungilike fruiting bodies. The fruiting bodies create haploid spores via meiosis, which then combine to become diploid and begin growing into amoebalike blobs when conditions again become favorable.

Acrasiomycota

Acrasiomycota are cellular slime molds that are haploid blobs that live independently in much the same fashion as other slime molds until food or water becomes unavailable. Interestingly, they then leave a chemical trail that acts like a *pheromone* to attract other cellular slime mold cells to a central location, where they form a *pseudoplasmodium*. During this stage, the individual cells, by still unknown means, communicate and coordinate their movements to act as one large blob. As the pseudoplasmodium moves, it can divide and (in unfavorable conditions, such as high heat or excessive dryness) form fungilike fruiting bodies that develop haploid spores. The haploid spores remain dormant until favorable conditions return, at which time they grow into individual haploid slime molds.

The Least You Need to Know

- The first organisms probably used RNA, not DNA, for growth and reproduction. Heterotrophs were likely the first organisms, but autotrophs became dominant.
- Simple forms of bacteria were probably the first prokaryotes; archaebacteria are similar to both prokaryotes and eukaryotes and are sometimes classified into their own kingdom.
- There is evidence that two successful invasions of a prokaryote cell created the eukaryote cell type.
- Endosymbiosis is the process that created internal membranes; membrane infolding assisted the prokaryotes invasion and may have created a second membrane for mitochondria and chloroplasts.
- Protists are very diverse. It is thought that some evolved into plants, others into animals.
- Slime molds exhibit two very different life-forms depending on the conditions.

Part 4

Plant Diversity and Systems Analysis

Now we enter the world of plants, which are a lot more like humans than most people think. For those of you who have ever struggled to keep a houseplant alive or who have failed to reap a bountiful harvest from your garden, read on. Everything you need to know about plant growth is included. We'll even cover some of the fun things that hormones cause plants to do and the mechanism that causes them to happen. The modification of simple plants into more complex organisms is examined through easy-to-follow steps that make sense and are easy to remember.

Chapter

Plant Systematics

In This Chapter

- Structural and nutritional requirements that support plant functions
- Methods that plants utilize to maintain their water balance
- Advantages vascular plants have over nonvascular plants
- How cell types affect tissue types in plants
- Xylem and phloem tissues within the vascular bundle
- Structural differences between roots, stems, and leaves

Plants have an interesting evolutionary history that demonstrates the ability of living things to adjust to a changing environment. All plants have basic requirements for life. When the requirements are met, life is good. The rest of the time, the plants have to struggle. Sometimes during that struggle, plants modify and become more fit for their environment and new species are formed; most of the time, they die. How successful are you with your houseplants or victory garden? I do not recommend stressing them to try to artificially create a new species; they may disappoint you and take the easy way out.

This chapter examines the requirements for a plant to survive and the modifications to plants' structures that have allowed them to process substances more efficiently. There is also a curious section on hormones; yes, they have them, too. Plant hormones regulate their growth in ways that maximize their survival potential, just like animals.

Evolution History

It is believed that over time, plants moved from aquatic territories to colonize the land. Unfortunately, the fossil record does not reveal much useful information about this time period because these early plants apparently were too soft-bodied to leave much of a print. Some fossil evidence is available indicating that certain algae could live out of the water as recently as 500 to 600 million years ago. The fossil record suggests that plants evolved from our modern-day algae, because both contain the same two chlorophyll pigments, have cell walls made of cellulose, and manufacture starch for stored energy. Compared to living in water, living on land presented two benefits: unfiltered sunlight, and wind currents that provided fresh carbon dioxide and removed the waste oxygen gas. Once out of water, plants had to develop several specialized structures to overcome the hardships of life on land, such as the following:

- Maintaining water balance
- Gaseous exchange
- Nutrient acquisition
- Exposure to sunlight
- Sexual reproduction

However, life on land also meant that water and nutrients were usually only available in the soil, and the dry atmosphere evaporated water from unprotected cells. Refer to the illustration *Plant phylogenetic tree* for a visual organizer of the evolution of land-based plants.

Maintaining Water Balance

When living in water, the problem of maintaining water balance did not exist. Because aquatic plants were also unicellular, they used osmosis to draw water into or out of the cells as needed. Once out of the water, plants faced two distinct water-related problems: getting water into the plant and keeping it inside. To satisfy these challenges, plants developed a vascular system made of specialized cells to transport water and food. Additionally, they created a waxy, lipid-based, hydrophobic cuticle to prevent water loss by evaporation to the dry environment. The cuticle prevented unwanted evaporation, but it also created an additional problem by preventing gaseous exchange.

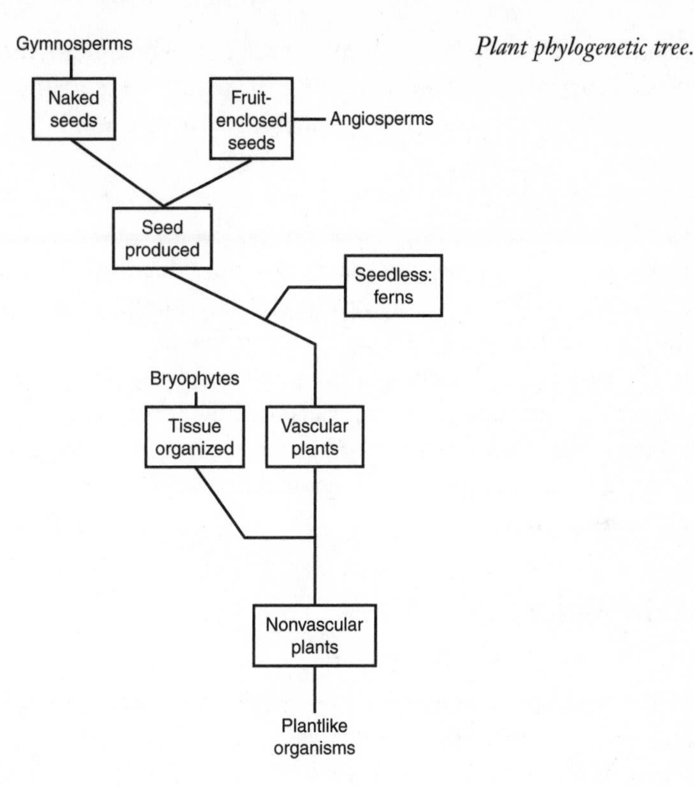

Plant phylogenetic tree.

Gaseous Exchange

Plants depend on free and unblocked gaseous exchange to support their primary life functions. They must absorb carbon dioxide from the atmosphere to perform photosynthesis and eliminate the waste gas, oxygen, to the atmosphere. To provide for photosynthesis, the waxy cuticle that prevents water loss is punctuated with specialized cells called *guard cells*, which bend to create openings in the cuticle membrane called *stomates*. When open, gases are free to diffuse through the membrane. Incidentally, the stomates are usually located in the shade on the underside of the leaf to further prevent unnecessary water loss.

Nutrient Acquisition

Land plants need to move a steady stream of water into the plant along with the raw material nutrients. Whereas the gaseous exchange normally takes place in the leaves, water and nutrients enter the plants through roots or rootlike structures. As plants moved onto land, they had neither, but fossil evidence suggests that a symbiotic relationship with fungi provided rootlike structures for absorption of water and nutrients in exchange for organic molecules, such as sugars, photosynthesized by the green plant. However, to photosynthesize, the plant had to maintain maximum solar exposure. Refer to the section "Typical Root Structure and Function."

Exposure to Sunlight

In the water, it was easy for plants to simply float on top to orient themselves in such a way as to receive solar energy. On land, shading by geologic features and airborne dust particles created the need for leaf orientation to continue photosynthesizing at the maximum level. Cellulose, a complex carbohydrate, is manufactured by the cell to serve as a rigid support in cell walls to allow for stem development and sturdy leaf support. For simple plants nothing more is required; more advanced plants utilize hormones (see the section "Hormonal Effects") to bend the stems and leaves for better solar reception.

Sexual Reproduction

Aquatic sexual reproduction is simple. Massive clouds of sperm are released by the males to travel on water currents, or swim to awaiting eggs, hopefully. The dryness of land prevents this type of exchange and demands a different mechanism. In response, plants developed a *gametangium*, which is a specialized cell designed to protect gametes, zygotes, and embryos from exposure to the drying and deadly effects of the environment. It is likely that the emerging land plants used gametangia as a nursery for continued protection of developing embryos. Land plants continued their specialized development to create separate lineages—one with vascular tissue that gave rise to more complex plants, called *tracheophytes*, and those without specialized conducting cells, called nonvascular plants (see Chapter 13).

Typical Plant Structure and Function

All living organisms are made of cells. Through the various evolutionary pathways, plants have developed three basic types of cells that are specialized to create all of the tissue required to support all plant structural and functional requirements. Each of the three specialized types has a protective cell wall, a semipermeable cell membrane, chloroplasts, vacuoles, and plastids in common.

Cell Types

The various functions of plants require structure for support, all of which are based on the following specialized cell types:

- *Collenchyma cells* have thickened cell walls and are usually arranged in strands to provide support for the plant. Collenchyma cells are especially designed for use by growing plants because their irregular shape and unconformity in the thickness of their cell wall can elongate as the plant grows.
- *Parenchyma cells* are the most abundant type of cells in plants because they have many functions. They serve as action sites during photosynthesis and aerobic respiration. They also have thin, flexible cell walls, which allow them to serve as water- and food-storage sites. Interestingly, parenchyma cells are responsible for the repair and healing of plant injuries because they have the unique ability to mitotically divide and differentiate into the other cell types as needed!
- *Sclerenchyma cells* have even-sided walls and an inner secondary wall hardened with lignin to provide structural support by the plant. Because they do not elongate like collenchyma cells, they are primarily found in the mature sections of plants. Sometimes sclerenchyma cells are organized into long, slender bundles called fibers, which historically have been used to make rope. Shorter, thicker, more irregular-shaped sclerid-type cells have more lignin, making them extremely hard, which makes them useful as seed coats. Incidentally, lignin is the hardening component in plants, and is the main ingredient in wood.

Bionote

The clothing fabric linen is made from the sclerenchyma cells of a flax plant.

Tissue Types

Similar cells working together to perform a specific function are called *tissue.* Plants need several different types of specialized tissue for construction and performance of the various system functions. Each of the four tissue types is constructed from the three cell types described in the previous section.

- *Meristematic tissue* is the location of continual growth and production of new cells by mitosis. Meristematic tissues divide and differentiate into four distinct tissue types: apical meristem, lateral meristem, vascular cambium, and cork cambium. The apical meristem is usually located in plant parts that are growing in length such as the roots and shoots (stems) called *primary growth.* Lateral meristems are also located near the outside of stems and roots and allow for growth in diameter, which is called *secondary growth.* Whereas the vascular cambium produces additional vascular tissues, cork cells are made from cork cambium and prevent water loss and provide insulating protection.

- *Epidermal* and surface tissue forms the outside layer of roots, stems, and leaves and is typically made of parenchyma cells. Some epidermal tissue specializes into threadlike root extensions to increase mycorrhizae associations and water absorption. Mycorrhiza is a mutualistic relationship between fungi and plant roots that increases water absorption. Others differentiate as guard cells that create openings called stomata for gaseous exchange, usually on the underside of leaves.
- *Ground tissue* or mesophyll is surrounded by epidermal tissue and is primarily composed of parenchyma cells that occupy the area between the vascular tissue and epidermis. Most of the cell functions, such as photosynthesis, metabolism, and storage, occur in the ground tissue. Most nonwoody plants, such as typical houseplants, are made of ground tissue.
- *Vascular tissue*, xylem and phloem, are major components of the vascular bundle. Both xylem and phloem consist of cells joined together to form tubes or pipes that provide a conduit for water and food to all parts of the plant.

Vascular Bundle: Xylem and Phloem Structure and Function

The xylem tissue is made of *tracheids* (which are found in all seed-producing plants) and *vessel elements* (which are found in angiosperms). Xylem forms the woody part of trees. Refer to the illustration *Typical xylem structure*.

Note that the structure of the xylem minimizes the backflow of water. If these structures were absent, it would be difficult to maintain a continuous water column to the top of tall plants. Loss of water-to-water contact would eliminate the cohesive property of water that holds water particles together and uniquely allows for uninterrupted flow.

Bioterms

Tracheids are elongated cells that allow water to enter and exit in small pores located on the tapers at each end. **Vessel elements** are wider and lose their cell wall where they are joined together to form a long, continuous tube.

Both tracheids and vessel elements mature, die, and dissolve their cell contents, leaving the structural cell wall to form the pipeline that transports water and minerals, only upward for plant use.

Phloem tissue transports a variety of substances, including sugars and nutrients, in all directions throughout the plant structure. The conducting parenchyma phloem cells have a cell wall, cell membrane, and cytoplasm. Other normal organelles are either dissolved at maturity or are modified into smaller structures. The main phloem cells are called sieve tubes, which are joined together like vessel elements. Nutrients and sugars move from the cytoplasm of the neighboring cell through tiny holes in the cell wall that act like a sieve, hence the name. Each sieve tube is surrounded by companion cells, which maintain their cellular organelles and are thought to carry out metabolic functions for the sieve tube cells, aid in the transport of food, and control the activities of the sieve tube cells. The phloem must be able to transport materials in all directions, so the backflow structures in xylem are not needed in phloem. Refer to the illustration *Typical phloem tissue.*

Bionote

Hemp, flax, and jute are sclerenchyma phloem fiber cells that have an economic benefit.

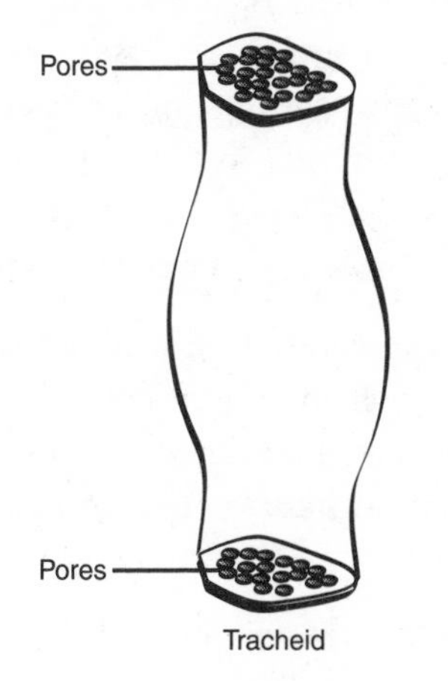

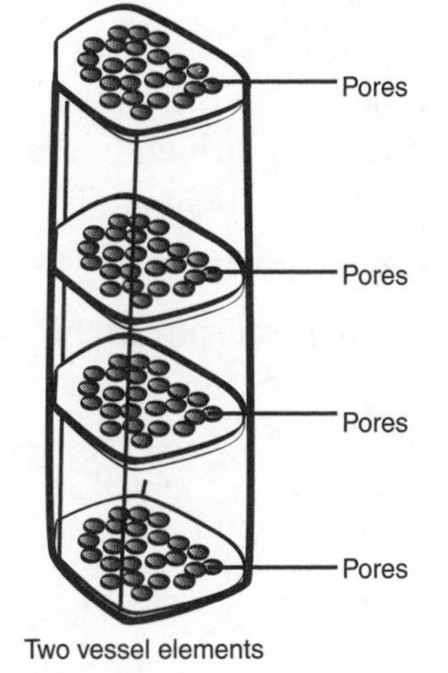

Typical xylem structure.

Sieve plate
Sieve tube
Companion cell
Sieve plate

Typical phloem tissue.

Typical Root Structure and Function

Whenever a seedling sprouts, it starts by producing a primary root, which begins the process of anchoring the plant; absorbing and transporting water, minerals, and nutrients; and storing organic compounds, such as sugar. Secondary roots follow and, in the case of most monocots, grow into a dominant, highly branched fibrous root system. In a dicot, the primary root usually remains the dominant root, which may grow to become long and thick, thus helping to anchor the plant. For your reference, a monocot is an angiosperm, or flowering plant whose embryo has a single seed leaf called a cotyledon; a dicot is an angiosperm with two cotyledons. In the case of most trees, the primary root can tap into underground water supplies.

The tissues that compose a mature root are classified into three types: epidermis, cortex, and vascular bundle. The epidermis in most plants is a single layer of flat cells that often create threadlike projections called root hairs, which greatly increase the surface area available to assist in the absorption of water and soil minerals. The cortex is ground tissue of mostly parenchyma cells that surrounds the vascular bundle and transports absorbed water and soil nutrients inward from the root hairs to the vascular bundle. The cortex region is the bulk of the root and is also used to store sugars or starches for reserved energy. The innermost part of the cortex is made of water-impermeable cells called the Casparian strip, which forces water and all nutrients to pass through that selectively permeable membrane to filter out unwanted substances. The vascular bundle is the innermost cylinder of the root, containing the xylem and phloem tissues.

The mechanism of drawing water from the soil into the root is different from drawing soil minerals into the root. Roots use osmosis to move water from a high-concentration area, such as normal soil, into the cell, which normally has lower water concentration because of the presence of amino acids, polysaccharides, and other organic compounds that dilute the water concentration.

The parenchyma cells composing the epidermis of the root hairs are permeable to water and dissolved inorganic ions (minerals). There are two mechanisms of getting them to the xylem from the root hairs.

Bionote

Water can be drawn out of a root back into the soil in a concentration reversal that happens whenever solutes, such as too much fertilizer, are added to the soil, making the water concentration in the soil less than inside the cell, causing *root burn* that may result in plant death.

In the first mechanism, the water is moved in an interior direction from the cytoplasm of one cell to the cytoplasm of the neighboring cell by specialized structures called *plasmodesmata*, which are channels through adjoining cell walls.

In the second route, the water and dissolved ions move into the cell walls of the root hairs and osmotically move to adjoining cell walls, never going through a cell, until encountering the impermeable Casparian strip, which forces the materials to go through a selectively permeable membrane before entering the xylem to keep out unwanted substances.

Because the concentration of ions is greater inside the cells of the root hairs, additional mineral ions have to be pumped into the cell by expending energy, ATP, in the process of active transport. This same ion concentration differential allows water to enter the root hair by osmosis. Once inside, the nutrients are pushed against the concentration gradient through the cortex until reaching the impermeable Casparian strip. Active transport is again needed to push the nutrients into the vascular bundle. As the concentration of nutrient ions increases inside the vascular bundle, water also osmotically moves in to equalize the concentrations. Because the nutrients cannot return through the Casparian strip, they continue to draw water into the vascular bundle. The resulting *root pressure* forces water into and up the xylem.

Root pressure generated by the buildup of water in the vascular bundle is not forceful enough to move the xylem sap to the top of tall trees and other large plants. As it turns out, the water is actually pulled up large plants by the interrelation of *transpiration* and cohesion. Transpiration is the continual evaporation of water out of leaves. This action then pulls adjoining molecules, because of their cohesion, to replace the molecules that were transpired from the plant. So the exit of one molecule of water from the top of the plant creates a ripple effect that terminates somewhere in the soil.

Bionote

The mineral nutrients most needed for plants are common in fertilizers: nitrogen, phosphorus, potassium, calcium, magnesium, and sulfur. In addition, trace amounts of seven other minerals are required.

It is estimated that 90 percent of the water absorbed by the root hairs is lost to the atmosphere by transpiration through the stomates. To control these openings, two opposing guard cells swell as osmotic water intake expands these C-shaped cells, creating a hole, the stomata, between two facing guard cells. When water osmotically leaves guard cells, they shrink, become more flaccid, and the stomates close.

An interesting mechanism governs the water flow into and out of guard cells, thereby adjusting stomatal opening and regulating water loss by evaporation. The presence of strong light or low carbon dioxide levels causes guard cells to actively transport potassium ions inside. The migration of potassium ions is followed by osmotic water movements trying to equilibrate the concentration. The increased water pressure expands the guard cells, creating the stomata. Gaseous exchange occurs normally as long as water loss by evaporation is replenished by water absorption from the environment so the water level is in balance. However, in times of drought, the water loss from the stomates may not be replenished completely and the plant wilts or, in extreme cases, dies. To prevent dehydration and possible death, guard cells close the

Bionote

During a drought, harvests are usually poor because the plant cannot produce the carbohydrates (especially sugar) that are the product of photosynthesis, because the plant cannot take in carbon dioxide. As a result, individual plants are considerably diminished in their stature and fruit-bearing capacity.

stomata. With the stomata closed, the plant cannot take in the carbon dioxide needed to undergo photosynthesis.

Organic compounds move via phloem in all directions in plants to the location where they are utilized or stored, in a process called *translocation*. It was once believed that phloem sap, containing a sugary solution of mineral ions, amino acids, and hormones, moved through the plant by diffusion. It was subsequently proven that diffusion alone would be too slow to support cellular functions. At this time, a pressure-flow mechanism is widely believed to power phloem sap movement.

The pressure-flow mechanism can be described in three steps:

1. The sugars are normally made in the mesophyll of the leaf and are actively transported into the phloem. Water follows by osmosis to equilibrate the concentration.
2. Sugar is continually translocated by the phloem for use at the processing end of the plant for metabolic functions or to be stored for later use. Again the water follows the sugar. As both sugar and water leave the phloem, it creates a low-concentration area that draws both sugar and water from the manufacturing site.
3. Because the water is not stored or utilized, excess water is recollected osmotically by xylem and returned to the leaves for reuse.

Sugars and other inorganic compounds are actively transported from photosynthetic sites in the mesophyll into the phloem sap stream. Water follows these substances by osmosis to equalize the concentration. Once in the phloem, the sugars are translocated to other areas, leaving behind a low concentration of sugars that helps draw more sugar and water into the phloem. Constant removal of sugar continually regenerates the concentration gradient. Excess water is absorbed by the xylem tissue and translocated to other areas.

Typical Stem Structure and Function

Stems function primarily to orient leaves toward a light source and to serve as conduits between the roots and leaves. However, some stems have specialized functions, such as the edible potato tubers (that serve as a food source), English ivy runners (that generate new plants), cacti stems (that store water and undergo photosynthesis), and devil's cane, which produces protective thorns (that deter touching). Stems are made of four basic tissue types:

- Parenchyma (pith)
- Vascular tissue (xylem, phloem)
- Cambium tissue (vascular and cork cambium)
- Cork tissue (bark)

Pith in monocots is used for storage and is interspersed throughout the stem. In dicots it is the center of the stem surrounded by xylem. Refer to the illustration *Typical stem structure.*

Vascular tissue is composed primarily of xylem, which after maturity becomes the non-functioning heartwood, and phloem, which is usually located underneath the bark. Pioneers often cut a band around a tree to disrupt the phloem translocation of organic substances so that the roots would die, followed by the rest of the tree; it was an easy way to clear a forest for planting. The vascular cambium separates the xylem and phloem and produces more of each, enabling the stem to grow.

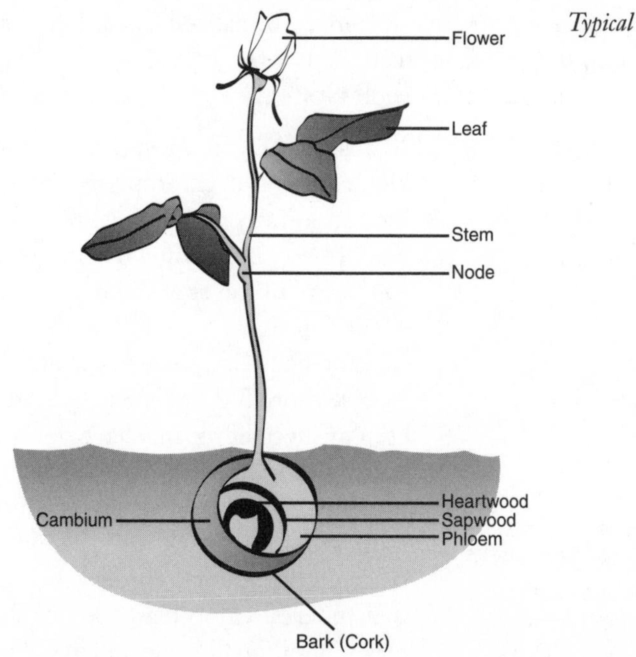

Typical stem structure.

Between the phloem and bark of woody dicot trees, the cork cambium produces cork tissue to create the waterproof bark that helps prevent water loss. The outermost cork cells that form the exterior protective bark are dead.

Stems are actually more structurally complex than roots. They contain growth segments called *internodes* that produce a *node* at each end. Nodes serve as attachment sites for leaves and also sometimes serve as rooting locations for artificial propagation. In addition, each node develops a lateral bud that contains meristematic tissue capable of developing into a new branch. Each bud is protected by a bud scale in a way similar to the root cap protecting a root tip, except the bud scale falls off when the bud begins to grow.

Typical Leaf Structure and Function

The primary purpose of leaves is to serve as a solar-collecting site to power photosynthesis, which makes their flat, green structures ideal for that purpose. They also serve as the gaseous exchange point, releasing oxygen and intaking carbon dioxide as a raw material for photosynthesis. An estimated 95 percent of a plant's dry weight is from organic compounds made from carbon dioxide.

Leaves exhibit more variation than stems. Specialized leaves function as animal traps, are modified into thorns for protection, as tendrils to wrap around objects for growth support, and are flattened into needles for environmental protection.

Leaves are structurally designed to absorb light. In most plants, photosynthesis takes place in the parenchyma cells, which contain the chloroplasts and are themselves located in the ground tissue of the leaf mesophyll. The leaves on the same tree are often differentiated so that small leaves at the top do not create overwhelming shade for leaves at the base. The reverse is also true in dry, desert-type areas where plants receive more light than they can use. In those areas, certain succulent plants have developed a fuzzy coating of specialized tissue to reduce light and provide shade! Whereas high-light leaves contain more chloroplasts per area than low-light leaves, the chloroplasts in low-light leaves orient themselves so that superior chloroplasts do not shade lower chloroplasts, additionally chloroplasts often move in a vertical Ferris wheel type of circle within the cell to give each chloroplast a solar blast on a rotating schedule.

Bionote

Carnivorous plants such as Venus flytraps, pitcher plants, and sundews, attract pollinators into their modified leaves, which serve as trapping devices that restrain the victim until their digestion is complete. The mineral nutrients received from the prey allow them to live in poor soil where other plants cannot survive.

Hormonal Effects

Both plants and animals manufacture hormones that function somewhere else in the organism. Hormones trigger enzyme activation, changes in membrane structure and function, and regulation of gene activity. Five types of hormones are produced in small amounts by the plants:

- Auxins
- Abscisic acid
- Gibberellins
- Ethylene
- Cytokinins

Auxins

Auxins are a type of chemical that causes cell elongation and are typically made in the meristematic region of the plant. The meristematic tissue produces a constant stream of a common, naturally occurring auxin, *indoleacetic acid* (*IAA*), which moves down all sides of a stem or branch. For a plant cell to bend, the cell wall must become flexible; IAA increases the pliability of the cell wall, allowing it to stretch. IAA appears to denature in light, so the shaded side of a stem retains greater concentration of IAA, which causes the cells to elongate more on the shaded side. When this happens, the plant stem bends toward the light, which is beneficial to the plant. This phenomenon is often noticed in window houseplants that do not get rotated. Refer to the illustration *Auxin effect.*

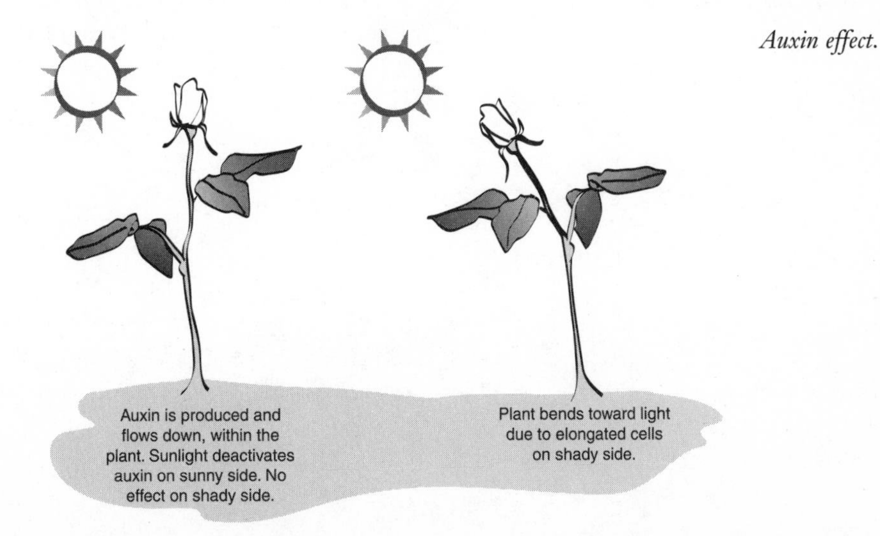

Auxin is produced and flows down, within the plant. Sunlight deactivates auxin on sunny side. No effect on shady side.
Plant bends toward light due to elongated cells on shady side.

Auxin effect.

Abscisic Acid

Abscisic acid (*ABA*) inhibits cell division, allowing a plant to remain dormant during harsh conditions, such as winter. ABA also causes seed dormancy until sufficient water is present to dilute the effect. ABA is antagonistic to the effects of gibberellins, which act to prevent dormancy. It is believed the ratio of ABA to gibberellin determines when a bud begins to grow.

Gibberellins

In 1926, Japanese scientists noticed that a substance produced by the gibberella fungus caused infected plants to grow abnormally tall. This active ingredient, named *gibberellin* or gibberellic acid, is also a hormone produced in small quantities in the meristematic tips of stems and roots, where it affects a variety of functions, including growth regulation. The main effect is the promotion of elongation in stems and leaves. As previously noted, it also works in concert with ABA to control bud dormancy. Further, in cooperation with auxins, gibberellins control fruit development and can make apples develop without fertilization. Interestingly, gibberellin sprayed on seedless grapes causes them to increase greatly in size, thereby increasing production and profit. Gibberellins are also used commercially to control seed germination so that a planted field will germinate at a given time to avoid the effects of herbicide application. Refer to the illustration *Gibberellin effect.*

Bionote

Gibberellins are added during the malting process of beer production to increase the sugar content. This in turn increases the alcohol content.

Gibberellin effect.

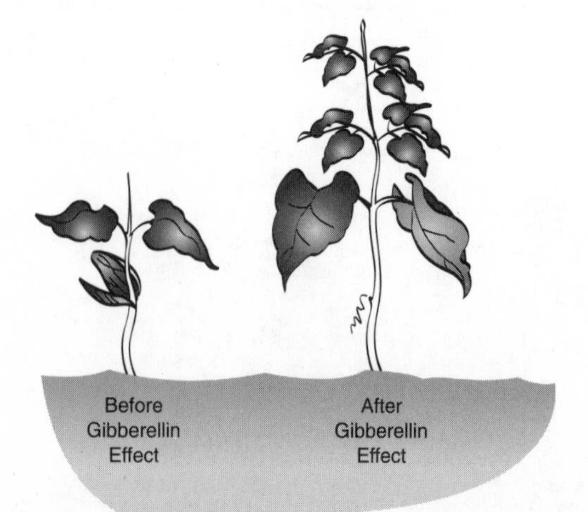

Ethylene

Ethylene is a gas at room temperature and functions to control the ripening of fruit, such as bananas, tomatoes, honeydew melons, and mangoes.

Bionote

Ethylene gas diffuses easily from one fruit to another, so the saying "one bad apple can spoil the whole barrel" is true and is related to the action of ethylene.

More specifically, ethylene controls the aging process. The ripening of fruit, like the autumn effects of color change and leaf drop, is all part of the aging process. Ethylene gas is produced normally by fruit to increase the effects of aging, such as cell wall breakdown, color change in fruit and leaves, and finally fruit and leaf drop. Leaf drop is an ecological advantage that works to rid the trees of unhealthy leaves and hasten their departure before cold weather and is based on a combination of auxin and ethylene interaction.

Bionote

Many growers spray their plants with cytokinins to increase branching. This is especially true of commercial Christmas tree growers.

Cytokinins

Cytokinins are produced almost everywhere in the actively growing plant tissue, including the roots, stems, leaves, fruit, seed, and function to promote cell division and retard the aging of flowers and fruit. Cytokinins work antagonistically with auxins. Whereas auxins work to make a plant grow taller vertically, cytokinins make the plant grow horizontally by increasing branching. The overall plant structure is regulated by the interaction of cytokinins and auxin. Refer to the illustration *Cytokinin effect.*

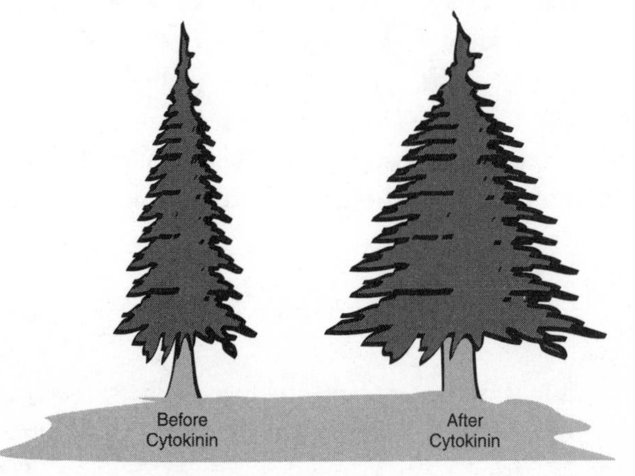

Cytokinin effect.

Plant Nutrition

Like you, plants require nutrients to grow, except their nutrients are inorganic compounds that we cannot metabolize. Plants take in carbon dioxide from the atmosphere, and water and minerals from the soil, and chemically combine these simple raw materials to make some of the most complex chemical compounds! To make these substances, plants have very specific nutritional requirements.

Requirements

In general, plants require large amounts of the following elements. There are 17 macronutrients; 5 of the most important are described next. The first three elements (nitrogen, phosphorus, and potassium) are listed in the order they normally appear on commercial fertilizers.

- *Nitrogen* promotes growth, especially in leaves. It is also a major component of plant protein such as nucleic acids.
- *Phosphorus* is needed in the development of flowers, roots, stems, and seeds. It is also used to make DNA.
- *Potassium* (potash), like phosphorus, is needed for the development of roots, stems, and flowers. It also helps to make proteins and carbohydrates. Potassium is also the main solute that drives the osmotic regulation in plants and activates several enzymes to catalyze reactions.
- *Calcium* combines with a protein to increase cell wall strength and helps cell membranes undergo cell division, metabolism, and growth.
- *Magnesium* is the central atom in the chlorophyll molecule and like potassium is a cofactor that activates several enzymes.

There are nine micronutrients: sulfur, iron, chlorine, manganese, nickel, zinc, molybdenum, copper, and boron. They are also called trace elements because plants only need small amounts. In fact, an overload of trace elements can be harmful for a plant. In most cases, the micronutrients function as cofactors for enzymes, similar to potassium and magnesium.

Deficiency Symptoms

When an element is required by a plant, and the plant fails to receive it in the amount needed, the plant exhibits deficiency symptoms for that nutrient. A wise gardener knows these signals and applies the appropriate nutrient before the condition becomes extreme:

- Nitrogen-deficient plants turn yellow and have stunted growth. This is the most common nutritional deficiency for plants.
- Phosphorus deficiency is almost as common as nitrogen deficiency. Insufficient phosphorus causes stunted growth, poor flowering, and a purple color on the underside of leaves.
- A potassium deficiency stunts root growth, which, in severe cases, affects stem and leaf growth. It is usually first noticed in older leaves that turn brown.

Nitrogen, phosphorus, and potassium are soil-borne macronutrients that critically influence the proper growth of plants.

Soil and Substrates

Soil is a combination of weathered rock and decayed organic material. The weathered rock is classified by size: Sand is the largest grain size, followed by silt, and clay is the smallest. Clay particles are so small that they can form an impervious water layer. In high-clay soils, flooding is more frequent not because of excess precipitation, but because the water does not soak into the soil very easily. It is difficult for sandy soils to hold water and for clay soils to drain water, so most plants prefer a silt-based soil type with sand and clay particles mixed in.

Decomposed vegetation and animal life create the nutrient-rich *humus* soil component. Humus is normally created on the surface of soil, where most plants receive their nutrition, and is often referred to as *topsoil* in combination with *loam.* Gardeners enhance productivity by plowing compost and green manure back into their soil to create more humus.

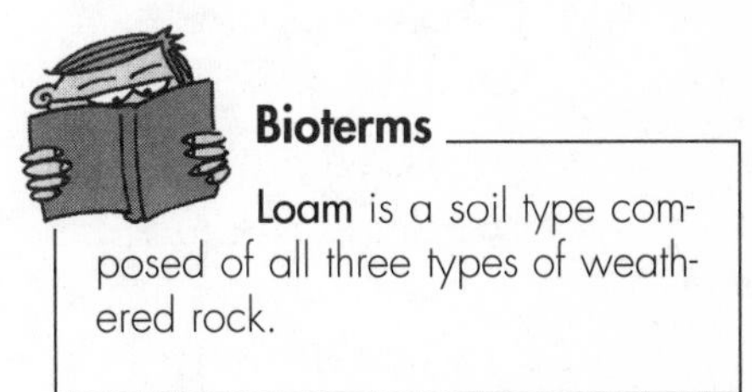

Bioterms

Loam is a soil type composed of all three types of weathered rock.

Compost is basically a nonmeat, vegetative material, such as grass clippings or kitchen waste, that is allowed to decompose. Green manure is a planted field cover that is plowed under to decompose beneath the soil surface.

Depending on the climate, it can take centuries to create a semifertile soil. Fibrous plant roots, such as grasses, help because they prevent erosion and the washing away of decomposed organic residue.

Plants exist within a narrow band of acceptable conditions. Certain plants have developed specialized structures that permit those plants to survive in fringe areas. Competition and herbivore action create natural-selection opportunities that determine the future of the population.

The Least You Need to Know

- Plants overcame numerous environmental challenges when first colonizing land by developing several specialized structural and functional mechanisms.
- Maintaining water balance continues to be an important concern for modern-day plants as it was for the first land plants.
- Plant functions require modified tissue and specialized cell types.
- Xylem and phloem tissues comprise the vascular bundle. They are structured to translocate substances needed by the plant.
- Plant hormones, like animal hormones, regulate certain cell activities to support the survival of plants.
- Plants have nutritional requirements; deficiency symptoms are noticeable if they are deprived.

Chapter 13

Plants and Plantlike Organisms

In This Chapter

- Determining whether fungi is a plant or animal
- Advantages vascular plants have over nonvascular plants
- Reproductive advantages for seedless and seed-producing vascular plants
- Life cycles of gymnosperms and angiosperms
- Dispersal and growth advantages of gymnosperms and angiosperms
- Why seeds are considered a reproductive advantage

Now it is time to talk about flowers. These wonderful creations are built for animal pollination, though it seems like for human enjoyment. Building on concepts learned from the previous chapter, this chapter deals with plants as whole entities. Several of the functions explained previously are put into play as the plants are examined.

Some of the plantlike organisms aren't really plants, depending on which kingdom classification to which you subscribe. However, they look like plants, they just don't photosynthesize. Fungi have been troublesome for taxonomists to classify between plant and nonplant since man found the first one.

Trees are much simpler. They are obviously plants. In this chapter, you will also learn about naked seeds and their productivity in certain areas. Their life cycles are presented in illustration and textual form to make your learning easier.

Nonvascular Plants

Nonvascular plants are the simplest type of plant. They are small; lack true roots, stems, and leaves; and have no specialized tissue for transporting water and dissolved nutrients. They are problems for taxonomists because they often do not fulfill the characteristics of a plant. Fungi are a perfect example. At one time, they were considered plants because they appeared to have roots. They are best described as a plant-like organism that is not photosynthetic.

Fungi

Fungi are eukaryotic heterotrophs that probably evolved about 300 million years ago. It is probable that they formed via endosymbiosis (refer to Chapter 2) as unicellular prokaryotes that remained joined together after mitosis, forming long chains of cells. They likely cooperated with the first aquatic plants to form a mutualistic relationship that allowed them to colonize land for the first time. Plant roots are often surrounded by fungal hyphae that absorb soil minerals, especially phosphorus, and transfer it to the plant roots. In exchange, the plant supplies products of photosynthesis to the fungus. This process, called *mycorrhizae*, produces an advantage for both organisms in semihospitable territories and enables them to survive in fringe environments, such as the original move from water to land.

Bioterms

A **hyphae** is a chain of cells joined together with either no cell wall or only partial cell walls between them, allowing the cytoplasm to flow between cells. They are also a component of sexual reproduction.

Fungi have several key evolutionary characteristics, including their reproductive style and a new twist on a life cycle. At one time, taxonomists classified fungi as plants because they appeared to have roots and reproduced by what looked like small seeds on recognizable reproductive structures. Today we know that about the only characteristic they have in common with plants is that they both grow in soil, or water, and are found almost everywhere.

- Most fungi, except for yeast, are multicellular and consist of long, interconnected filaments called *hyphae*, which continue to grow into a meshlike mat called *mycelium*.

- Nuclear mitosis in fungi is different from other eukaryotic cells because the chromosomes division occurs in the nucleus, not the cytoplasm of the cell. Nuclear mitosis ends when the nuclear membrane divides the nucleus into two separate nuclei.
- Fungi contain the polysaccharide chitin in their cell walls instead of cellulose, like plants. Chitin is also the hard protective shell used by insects.
- Fungi are not mobile like slime molds or animals. Instead, their mycelium grows so rapidly in search of food and water that the need for mobility is minimized.

Except for yeast, which reproduces by mitotic cell division, most fungi have three distinct stages in their life cycle, including a dikaryotic stage during which mycelial cells contain two haploid nuclei from two different parents. Interestingly, after fertilization from the dikaryotic stage, the diploid zygote undergoes meiosis to produce haploid spores. Upon finding favorable conditions (for example, adequate food and water supplies), the spores grow their hyphae into mycelia. They sometimes grow next to and connect with mycelia from different yet compatible fungi. The genetic contents of the hyphae transfer to the neighboring hyphae. These conjoined hyphae now contain two different nuclei in one fused cell (dikaryote), which continues to grow and reproduce as a dikaryotic organism. Finally, the dikaryotic mycelia produce a fruiting body, which is usually the only above-ground event; the common mushroom or toadstool is the observable fruiting of the dikaryotic stage. Within the toadstool, the nuclei combine to produce zygotes, which are then released to the environment. The zygote is the only diploid stage; spores are haploid; and mycelia and fruiting bodies are dikaryotic (the dikaryotic stage is interesting because it refers to the situation where cells contain two nuclei). Refer to the illustration *Typical fungi life cycle.*

Typical fungi life cycle.

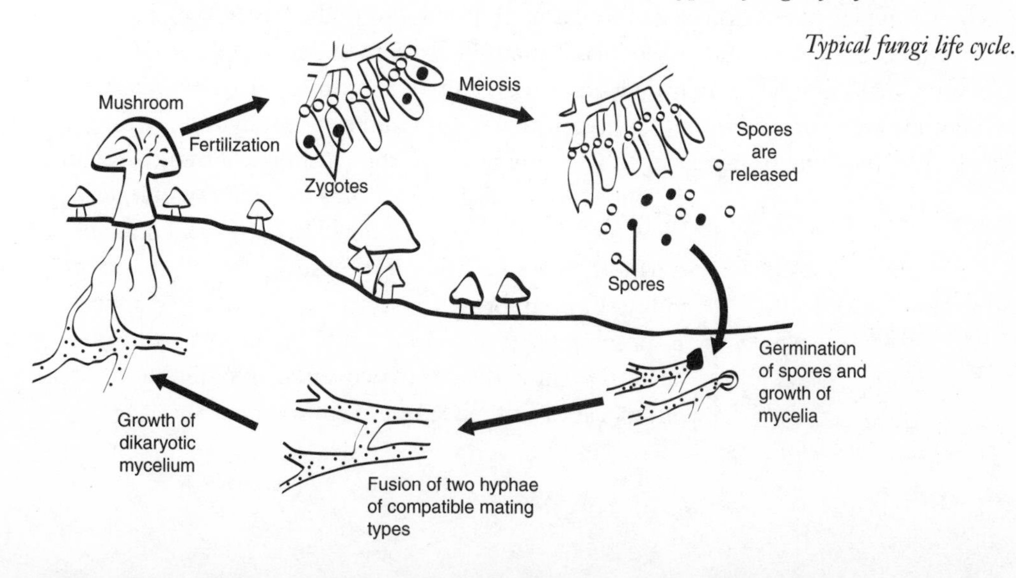

Bryophytes

Bryophytes are specialized plants that contain some of the cells needed for land colonization. Although they have a protective cuticle and their embryos developed in gametangia, they lack true roots, vascular tissue, and extensive cellulose support. As a result, bryophytes are short (usually less than 2 inches or 5 centimeters tall), only thrive in moist environments for simplified water intake, and use water as a facilitator of sperm mobility in sexual reproduction.

Like algae, they undergo an alternation-of-generations life cycle, alternating between a haploid gametophyte and diploid sporophyte stage. The gamete is the most noticeable stage, with the smaller sporophyte stage growing on the gametophyte, usually as a stalk with a capsule of spores on top. The gametophyte stage also produces rhizomes, which are threadlike cellular extensions that serve as holdfasts or anchors that keep the plant attached to its growing surface. Mosses, liverworts, and hornworts are typical bryophytes. The velvety green mat of mosses, usually seen in the dense shade, is actually the gametophyte stage.

Seedless Vascular Plants

Ferns are the typical seedless vascular plants. There are more than 10,000 different species today that may have evolved beginning 400 million years ago. They are typical tracheophytes with complete vascular systems, roots, leaves (fronds), and rhizomes, which are underground stems capable of producing new plants. Ferns grow throughout the world in moist temperature or tropical areas. They produce spores in sporangia on the underside of their fronds.

Seedless vascular plants dominated Earth until about 200 million years ago and are considered the first land-based "true" plant because they developed complex structures that eliminated their dependence on aquatic territories. Two interesting adaptations are thick-walled, drought-resistant spores and a sporophyte stage that is longer than the gametophyte stage. The enhanced sporophyte stage allows the plant an opportunity to utilize photosynthetic capabilities. This modification created new life cycles and opened new territories for colonization for green plants by providing more photosynthetic area and increased spore-disposal opportunities. Probably the most significant advancement is the organized vascular system because it allows easier translocation of substances.

Bionote

The first vascular plants did not produce seeds, but relied on the production of spores.

Vascular tissue consists of two types of specialized cells, xylem and phloem. Xylem tissue moves water, and phloem translocates the raw materials and final products of photosynthesis to every cell in the plant. Xylem tissue uses the cohesive nature and surface tension of water to move water and dissolved solutes independently of the phloem, which moves organic nutrients and manufactured products. The ability to circulate water and organic molecules plus the lignin-enhanced xylem tissue enables vascular plants to grow taller than nonvascular plants and also grow true roots, stems, and leaves.

Xylem tissues grow, mature, and die before they become water-transport tubes. They also only carry water upward; conversely, phloem tissue is still alive and conducts substances in all directions.

Seed-Producing Vascular Plants

Ancient plants developed extensive vascular systems, true roots, stems, and leaves as part of their conquest of land. To this list of adaptations, plants added a means of protecting the zygotes from temperature extremes and moisture availability, called *seeds.* A seed is a survival capsule that contains a tiny dormant embryo and a large food supply for a new plant enclosed in a tough seed coat that protects the contents from harsh environment conditions.

Embryos may remain dormant in seeds for prolonged periods of time until environmental conditions favor their emergence and growth. Normally water entering the embryo area through the seed coat triggers an end of dormancy. With the capability of seed production, land plants no longer needed water for transportation of sperm; instead they produce pollen, which transfers the sperm-producing cells to the eggs. In a sense, seeds are mobile, reproductive capsules.

Seed-producing plants are able to colonize territories inhospitable for seedless vascular plants, which open tremendous new areas for their dispersal. When a seed breaks dormancy and begins growing, the process called *germination.* After germinating, the embryo grows into a seedling. From here, it can grow into one of the two broad groups of plants. Either they produce seeds with no fruit—gymnosperm, or seeds inside of fruit—angiosperm.

Gymnosperms

All gymnosperms produce seeds but not seeds enclosed in a fruit. Instead, they produce "naked" seeds, which is the definition for gymnosperm. They are the oldest lineage of modern-day seed plants. Interestingly, their leaves have sometimes evolved into specialized reproductive structures called *scales*, both male and female, which are collectively part of a larger structure called *cones*. Cones are common features on pine trees. Male cones produce the male gametophytes called pollen; female cones produce the female gametophytes called eggs. After pollination, the female cone contains the seeds that develop on their scales. Clouds of tiny pollen grains are usually carried by natural forces such as wind and animals, allowing pollination to occur over a wide dispersal area and in very dry territories.

Bionote

Pollination is the process in which pollen is transported to the stigma or carpel of the female plant. Fertilization occurs after pollination.

Conifers, Cycads, Ginkgo, and Gnetophytes

The four types of gymnosperms do not vary greatly from each other. The conifers are the most successful, abundant, and representative of the gymnosperm plants. Conifers are also called evergreens because their leaves have been modified to needles, which do not defoliate at the onset of shorter, colder days like deciduous trees, but are replaced continuously regardless of the season.

Pine, spruce, cedar, redwoods, and yews are all conifers that trace their lineage to the Devonian period, about 400 million years ago. Cycads produce long, beautiful, palmlike leaves on short stalks and trace their history to the Triassic period 225 million years ago when dinosaurs roamed the earth. Cycads are endangered because of their slow growth rate, habitat destruction, and over-collecting for profitable purposes as ornamental plants. Native to the tropics, they are either male or the female plants, both of which develop large, showy, cones.

Bionote

Bald cypress and larch trees usually lose their leaves each fall. However, on some of these trees, the leaves are not lost and stay active on the trees up to 12 to 14 years!

Ginkgo biloba is the only ginkgo that has survived to modern times. It is also considered a living fossil because it bears a remarkable resemblance to the fossil ginkgoes from 125 million years ago. Its survival may have been a direct result of its fleshy edible seeds, which are considered delicacies in Japan and China where the trees have been cultivated for thousands of years. Ginkgo trees are either male or female; however, their most unusual characteristic is their deciduous trait of dropping their leaves in the autumn season.

Gnetophytes are a diverse group of plant structurally that includes not only trees but also shrubs and vines that produce pollen and seeds in cones that look like flowers.

One interesting species of shrub is the ephedra, which is common in the American Southwest, where it is also called Mormon's tea because it can be brewed into a tea to make the decongestant ephedrine.

Life Cycle of a Conifer

The life cycle of a pine tree is typical of the gymnosperms and demonstrates the reversal of prominence of the haploid and diploid generations as plants moved onto land. Refer to the illustration *Typical pine tree life cycle* for visual reference.

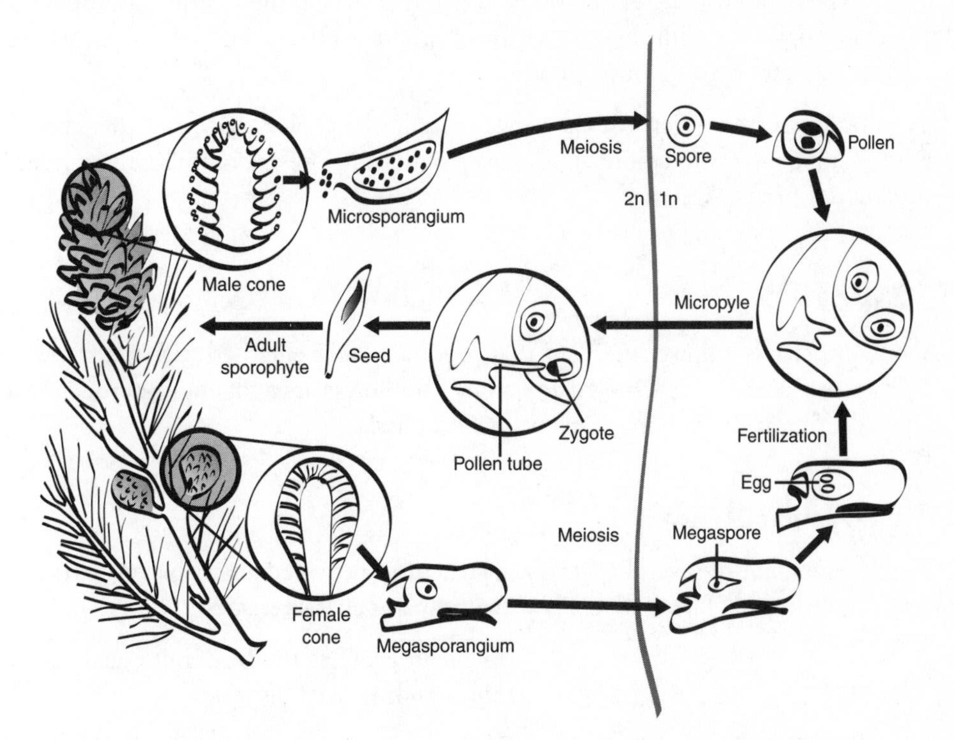

Typical pine tree life cycle.

Conifers, like the other gymnosperms, produce both a male and a female cone. The tree itself is the sporophyte stage, whereas the gametophyte stage is reduced to microscopic male and female gametes that develop in their respective cones. As such, most of the life cycle is spent in the diploid stage and can be described in four steps:

1. Each spring, clouds of haploid pollen grains are released into the atmosphere by the male pollen cones, after undergoing meiosis to land on the sticky female seed cones, in the process called pollination.

2. The pollen grain produces sperm cells by mitosis. These sperm cells then leave to fertilize the haploid ovules to produce a diploid zygote.
3. The zygote grows into a diploid sporophyte embryo as the entire ovule develops into a seed, which can then be released to disperse and grow in favorable conditions.
4. The diploid seed grows into a mature tree that produces either male or female gametes in cones, and the process repeats.

In this type of alternating generations, the pine tree is in the diploid, sporophyte stage most of its life with the exceptions of brief periods during the production of haploid, gametophyte sperm, and eggs.

Gymnosperms create tremendous forest areas in the cooler territories that are difficult for angiosperms to colonize. Massive coniferous forests dominate the mountainous areas of North America.

Angiosperms

An angiosperm is a flowering plant that produces seeds enclosed within a fruit. The fossil records indicate that approximately 90 million years ago, angiosperms replaced gymnosperms as the dominant plant type on Earth.

Evolutionary Advantages

The development of several specialized structures provided angiosperms with a decided advantage. There are six key evolutionary characteristics:

- The fruit protects the seed, but equally important, it aids in seed dispersal.
- The reproductive cycle in angiosperms can be simplified to seed-plant-seed and can occur in one year as opposed to multiple years for the gymnosperm life cycle. The ability to reproduce faster creates new territories and enhanced genetic-variability opportunities at the expense of the gymnosperms.
- Mycorrhizae associations with fungi are more pronounced in angiosperms, producing a more efficient vascular system.

Bionote

In addition to animals digesting the fruit and expelling the unharmed seed intact with an accompanying fertilizer, some plants have developed a specialized mechanism that launches the seeds away from the plant when they are burning.

- Animal coevolution has provided specific animal pollinators, which increases the chance of pollination beyond wind-assisted methods.
- The diversity of angiosperms is expressed in the approximately 230,000 known species. The variability is an advantage in inhabiting new territories and lifestyles.
- The production of seeds allows for delayed germination and growth. This provides offspring the opportunity to wait for favorable growth conditions when life expectancy increases. Seeds also contain an endosperm that is a supply of stored food for the developing embryos.

Monocots and Dicots

There are two main types of angiosperms: monocots and dicots. A monocot has one cotyledon (seed leaf), and a dicot has two cotyledons in their embryo, hence its name.

Monocots and dicots are characterized by also having four different plant structures:

- Monocots have leaf structures in which the veins run parallel to the main leaf vein; dicot veins are branched.
- Within the stem, the arrangement of vascular bundles in monocots varies, but in dicots, the various bundles are organized in a ring close to the bark.
- Flower parts, such as petals, are found in multiples of three in monocots and multiples of four or five in dicots.
- Monocots have branching root systems, which allows them to prevent erosion; dicots have long, straight taproots that usually break off when you try to pull them, and then grow back (dandelions, for instance).

Typical monocots include bananas, bamboo, corn, lilies, tulips, oats, onions, rice, and wheat. Typical dicots are beans, cacti, carnations, lettuce, roses, and most deciduous trees.

Typical Angiosperm Life Cycle

The life cycle of an angiosperm shares commonality with gymnosperms. Like gymnosperms, angiosperms spend most of their life in the diploid, sporophyte stage. The angiosperm life cycle can be described in three steps:

1. The production of gametes begins with meiosis in the specialized structures called anther for sperm and ovules for eggs.
2. Haploid gametes form and then grow mitotically to form mature gametophytes that produce the haploid gametes.
3. The male and female haploid gametes combine, fertilization takes place, and the resulting diploid zygote grows into a mature diploid sporophyte plant that then produces gametes as in step 1.

Refer to the illustration *Typical angiosperm life cycle* as an accompaniment to the text.

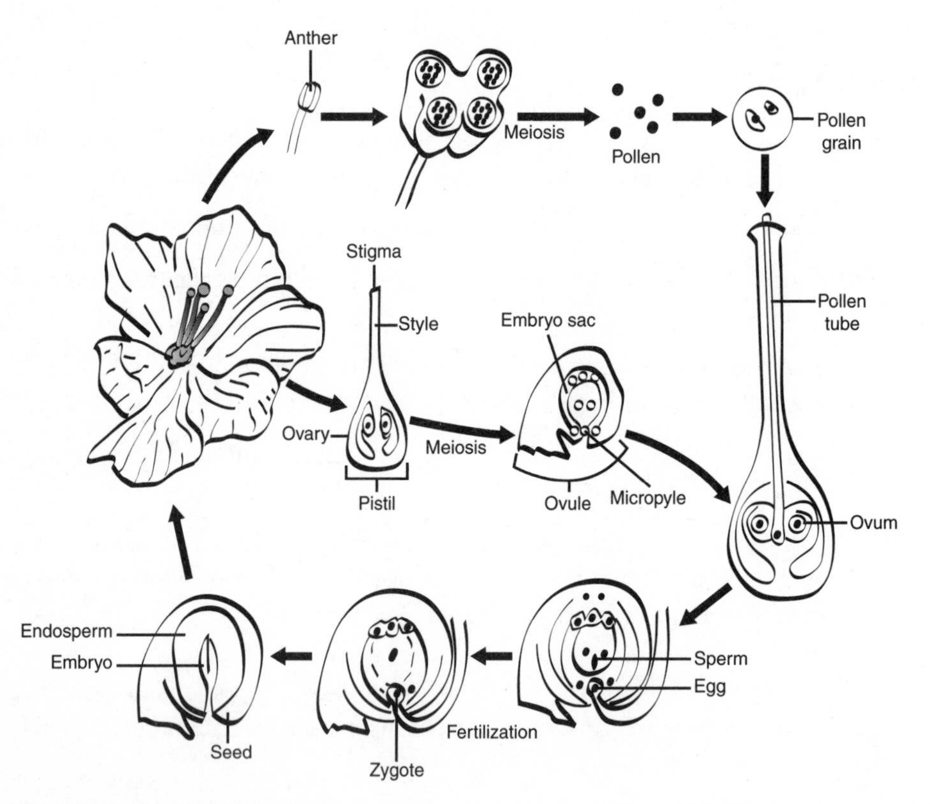

Typical Angiosperm Life Cycle.

Typical Angiosperm Reproductive Cycle

Angiosperms complete a *double fertilization* during their reproductive cycle that provides an advantage for their offspring. This unique cycle is described in the next steps:

1. The ovaries of a typical flower contain many ovules, each of which has a central cell that undergoes meiosis to produce four haploid spores. As in oogenesis, three degenerate and one divides mitotically to produce the female gametophyte, an embryo sac.

Bioterms

The process whereby one sperm unites with an egg to form a zygote while the other sperm of the pair unites with the central cell to become endosperm is called **double fertilization**.

2. The anther undergoes meiosis to produce four haploid spores, each of which then divides mitotically to produce a tube cell and a generative cell. A protective covering forms around both cells to create a mature pollen grain.
3. Pollination occurs when the pollen arrives at the stigma, where each pollen grain germinates and the tube cell develops into a pollen tube, which then grows to join the ovary. At the same time, the generative cell divides mitotically to form two sperm.
4. The pollen tube releases both sperm for fertilization of the egg. One sperm fertilizes the egg, while the other haploid sperm (N) joins with the diploid central cell (2N) to become a triploid (3N) cell. This mega-cell develops into the endosperm that provides a food source for the developing embryo.
5. The embryo divides mitotically to become a mature seed surrounded by a tough seed coat or shell.
6. While the embryo is developing, the flower degenerates and the ovaries divide mitotically to become the fruit.
7. At this time, metabolic activities rest as the seed enters dormancy, a temporary stage that allows time for seed dispersal and favorable growing conditions. The ovaries become the fruit; the ovules, with the help of the sperm, develop into the seed, usually within the fruit.

Refer the illustration *Typical flower anatomy* for a visual reference.

The simplest plants are single-celled organisms that contain chloroplasts. As design complexity increased, plants were able to colonize the land. In so doing, they developed true roots, stems, and leaves to provide the machinery necessary for photosynthesis to occur. Advancements in the vascular system and water-retention capabilities allowed greater colonization of dry environments. The development of seeds completed the transition from aquatic to terrestrial environments.

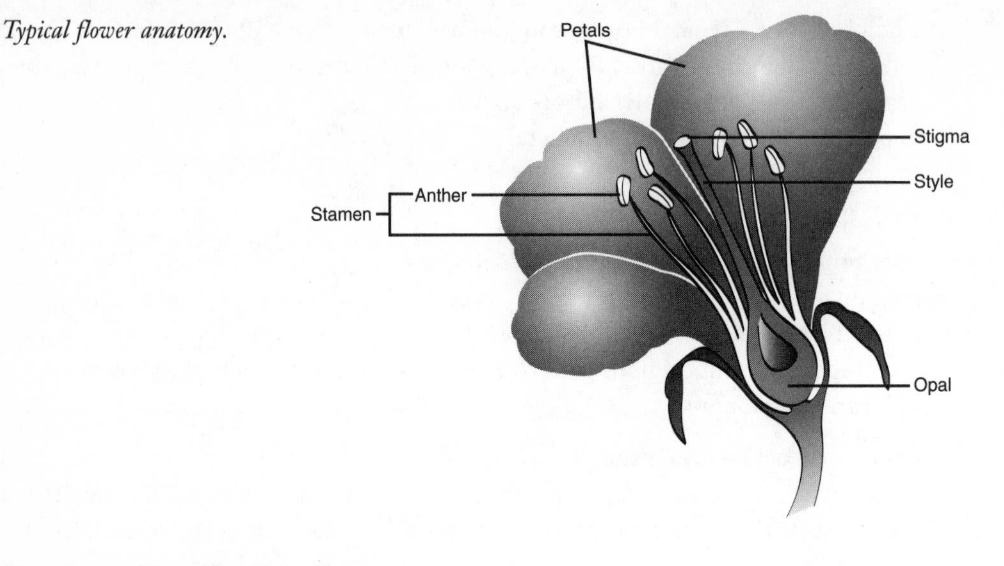

Typical flower anatomy.

The Least You Need to Know

- Fungi are common nonphotosynthetic eukaryotes that secrete enzymes to digest food outside the body.
- Maintaining water balance continues to be an important concern for modern-day plants, as it was for the first land plants.
- Seedless vascular plants emphasized the gametophyte stage of alternating generations, whereas seed-producing vascular plants emphasized the sporophyte life cycle.
- The dominant sporophyte stage created greater opportunity for plants to photosynthesize raw materials into usable products.
- Both gymnosperms and angiosperms have developed unique identifiable characteristics and different life cycles that enable them to survive.
- Plant functions require modified tissues and specialized cell types.

Part 5

Animal Diversity

This is the part that explains how your body works, or fails to work. Each system is presented in a way that demonstrates the advancements in that system from simpler animals to more advanced animals, with an emphasis on human features. The reproduction section has a PG rating because it explains the conception process from both male and female anatomical and physiological perspectives. How would you know if you or one of your friends is a Cro-Magnon or Neanderthal? Read on—there is an entire section that develops human history based on fossil finds from archaeological digs. Find out for yourself and then look in a mirror.

I DON'T CARE WHAT THE BIOLOGIST SAID. OUR HEARTS ARE NOT SIMILAR.
RJKING

Chapter 14

Digestive and Excretory Systems

In This Chapter

- How organ systems coordinate to maintain homeostasis
- Structural characteristics unique to vertebrates
- How nephrons eliminate nitrogen but conserve water and salt
- Tissue modification for increasing digestion efficiency
- Similarities of the human and frog digestive systems
- An alimentary canal and digestion

This chapter starts a new section you should like because it is the first in a series that explains how your body functions. It also compares your body with an evolutionary eye on how simpler animals did the same thing with lesser equipment. So why do we need more?

Essentially, the digestive system breaks down food into small enough pieces that they can be absorbed by the circulatory system. If you can understand that much, the rest is even easier, just longer.

Likewise, the excretory system is related to the digestive system. The excretory system handles metabolic wastes and maintains the water balance for the organism. The removal of harmful metabolic wastes has been a function of animals since they were all aquatic. When they moved onto land, the excretory system added the maintenance of an appropriate water balance to their list of functions. Simple animals do it, too; they employ the same biological principles you enjoy.

The Process of Digestion

Digestion is the process whereby food that has been ingested is broken down by mechanical and chemical means into a form that is usable by the organism. The *digestive system* is the series of specialized cells, tissues, and organs that accomplish the digestive process. The increasing length and complexity of the digestive system parallels the advancement in complexity of animals. In general, unicellular animals absorb their food and digest it within the cell. Multicellular animals exhibit a division of labor such that the food is digested outside the cells and is then delivered to the cells, usually by the circulatory system.

Invertebrate Digestive System

Invertebrate digestive systems are deceptively simple. In the simplest invertebrates, the food is absorbed and digested by each individual cell. However, even in this simple mechanism, a system of receptors and triggering devices must alert the cell to the presence of a potential food source and orient the cell for ingestion. Not much is known about this recognition and reception mechanism except that it appears to be a protein-to-protein reaction that causes a change of shape in the membrane structure, which permits and promotes penetration. The digestive enzymes produced by the cells are also a mystery. It is known that they are types of hydrolytic enzymes that break down food by removing water, but little is known about the first cell that successfully created digestive enzymes!

Porifera

Porifera, like sponges, and unicellular animals absorb and digest their food inside of the cells. This is the most efficient form of ingestion and digestion. However, it limits the three-dimensional depth of an organism to two layers because every cell must have access to the environment. It also limits the size of the food because the cell membrane cannot absorb something that is larger than it can stretch.

Cnidarians

Cnidarians such as jellyfish, corals, and hydras are the first animals to digest their food extracellularly. Food is captured and ingested into a gastrovascular cavity, a hollow space lined with specialized endoderm cells that secrete enzymes to chemically digest food into smaller pieces. The gastrovascular cavity only has one opening, so the entry is also the exit. Although the gastrovascular cavity allows an organism to utilize a food type physically larger than its cells, the one opening into the digestive apparatus limits its efficiency. When the extracellular digestion is completed, specialized gastroderm cells lining the cavity absorb the nutrients and digest them even further. When the intracellular digestion is complete, the waste materials are ejected back into the gastrovascular cavity for passive transmission back into the environment. Nutrients in the gastroderm are diffused to the remaining cells. Cnidarians do not have a circulatory system that delivers nutrients and removes wastes from each cell, so the process of diffusion not only limits the efficiency of this process, it also limits their size.

Bionote

Some cnidarians have symbiotic, photosynthetic protists inside the cells of their gastroderm. These autotrophs recycle the metabolic wastes of the heterotrophic cnidarians to produce sugars and other organic compounds that are then used by the cnidarians. This relationship is so important to the cnidarian that they prefer to inhabit light areas to support the photosynthetic capabilities of the protists!

Platyhelminthes Digestive Systems

Platyhelminthes is the phylum of flatworms (see Chapter 20). They exhibit both a parasitic and free-living species. The digestive systems are vastly different because of their different lifestyles.

Parasitic Flatworm Digestive System

The parasitic flatworms have much simpler digestive requirements than the free-living species. Because most parasites live off the blood or body fluids of their hosts, their digestive system does not need to be extensive or complex. For instance, tapeworms live attached to the intestinal wall of their human host and tap into the nutrient-rich blood. Because the food is already digested, they merely absorb nutrients from the host's blood into their bodies. There is no need for them to have a digestive system! In other parasitic flatworms, the muscular pharynx sucks blood or body fluid from the host into one-way intestinal storage areas where the digestive enzymes complete the digestive process. Parasitic lifestyle has an abbreviated digestive system because the work of digestion is completed by the host.

Free-Living Flatworm Digestive System

The digestive system of free-living flatworms is more complicated than that of the parasite. The free-living worms are scavengers on decomposing organic matter and on smaller organisms such as protists. Their muscular, tubelike pharynx extends from the opening to the highly branched gastrovascular system and is therefore able to transport food from the environmental to the digestive areas. Digestive enzymes in the gastrovascular cavity digest the food, and the nutrients are absorbed by the cells lining the system. Like cnidarians, the food is then diffused to the remainder of the worm. Waste and nondigestible foods are returned to the environment in a reverse trip through the pharynx. Free-living organisms must include the cells that create and secrete digestive enzymes and the location for the process to occur as part of their anatomy.

Nematoda Digestive Systems

Roundworms are long, threadlike organisms (see Chapter 20). Their digestive systems are similar—long and slender with an opening at both ends. Nematodes are the first animal group to have an *alimentary canal*, which is considered a complete digestive system because of the efficiency related to food entering an anterior mouthlike area and undigestible wastes leaving through a separate, posterior anal opening. This is a significant digestive advantage that is expressed more completely in the digestive systems of more advanced animals. Because the food is only moving in one general direction, different areas of the digestive system can be specialized for the digestion of different foodstuffs. This allows for a diversification in the types of food sources an animal can consume and receive nourishment from. It also means that a section may specialize in the manufacture and secretion of specific digestive enzymes and another section may specialize in the absorption of nutrients.

Free-living roundworms ingest almost anything from plantlike food sources such as algae and fungi to smaller protists and even decaying organisms. Parasitic roundworms are similar to parasitic flatworms in their preference for blood and body fluids. The unique presence of a unidirectional alimentary canal sets the stage for greater advancements in more complex animals.

Annelida Digestive System

Annelida is the phylum containing the common earthworm. The digestive tract of the annelids is a long tube that extends the length of the body from the most anterior structure, the head, to the most posterior structure, the anus. Structurally, the digestive system is like a tube within a tube. The digestive tube has several specialized structures that improve digestion. Refer to the illustration *Earthworm digestive system.*

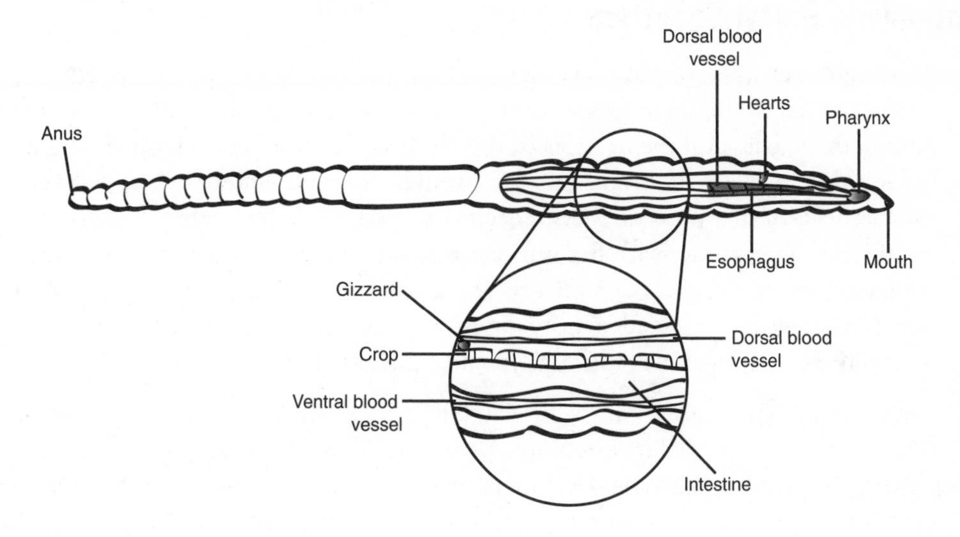

Earthworm digestive system.

This digestive plan is carried forward into more advanced animals. For the annelids, the pharynx, crop, and gizzard are notable modifications.

The pharynx of certain annelids is even able to project out of the mouth to capture prey. In carnivorous annelids, the pharynx also has two or more jawlike structures that are used to grab the prey. Herbivorous annelids use their modified pharynx to tear off chunks of vegetation for consumption.

The crop and gizzard are somewhat similar in structure and function. Both are saclike enlargements of the digestive system that can store food for later processing. The gizzard also has muscular connections that allow it to churn the contents. In annelids, the gizzard normally contains small particles of rock or sand to assist in the mechanical crushing of the food.

The common earthworm exhibits typical annelid digestive functions. As they tunnel through the soil, the pharynx ingests basically everything it comes in contact with, including the soil itself. As the soil contents move through the esophagus, it is usually stored in the crop and moistened with lubricants and digestive enzymes until the gizzard is physically able to handle more contents. In times of heavy feasting, the crop may bulge (*crop* is Greek for "bulge") before emptying into the gizzard. After the gizzard grinds up the contents, the material is moved to the lengthy intestinal area, where the nutrients are absorbed through the cell lining and into the bloodstream. Unusable material keeps moving and eventually is returned to the environment via the anus.

Echinoderm Digestive System

The echinoderms add an interesting twist to digestion. As the tube feet (arms) of a starfish attach to the shells of a clam or other two-shelled mollusk, they continually pull on the shells until the muscles of the mollusk tire and the shell opens slightly. The starfish then projects its *cardiac stomach* through its mouth and inserts it into the shell of the mollusk. In that process, the stomach is turned inside out so the digestive enzymes are therefore in contact with the soft organ tissue of the mollusk. After the digestion is complete, the cardiac stomach retracts and empties the contents into a second stomach, the *pyloric stomach*, which adds more digestive enzymes to complete the process. The nutrients are absorbed into the coelom and transferred to the rest of the body.

Invertebrate digestive systems begin with simple unicellular capture and internal digestion. More sophisticated invertebrates have specialized tissues that mechanically and chemically digest foods as well as other structures that specialize in absorption of nutrients. The advancement of an alimentary canal to replace the gastrovascular system is major. Most of the digestive structures are continued in the vertebrates.

Vertebrate Digestive Systems

The tissue specialization started by invertebrates is continued and elaborated on by the vertebrates. The movement toward a greater division of labor is characteristic of increasing animal complexity.

Fish Digestive System

Many fish are omnivorous, meaning they eat both plant and animal food sources. This requires them to have an elaborate digestive system. Fish have a true mouth with moveable jaws and teeth. Some species have a more flattened type of teeth for crushing food, whereas others have sharper teeth, often pointed inward to prevent prey from escaping. Even though they possess teeth, most fish normally tear off chunks of food or swallow it whole. The chunks pass through the esophagus and are collected in the stomach, where stomach acid and digestive enzymes continue the food breakdown. Some species of fish also have a *pyloric cecum*, which is a fingerlike pouch be-tween the stomach and intestine. The pyloric cecum also secretes digestive enzymes to continue digestion and nutrient absorption. After the partially digested food is moved to the intestine, digestive enzymes from outside the organs contribute enzymes to the process. Fish are the first animals to add hydrolytic enzymes from the pancreas to assist in the chemical digestion of food. The liver secretes bile to emulsify fats, and the pancreas releases protein, carbohydrate, and fat-digesting enzymes. Refer to the illustration *Fish digestive system.*

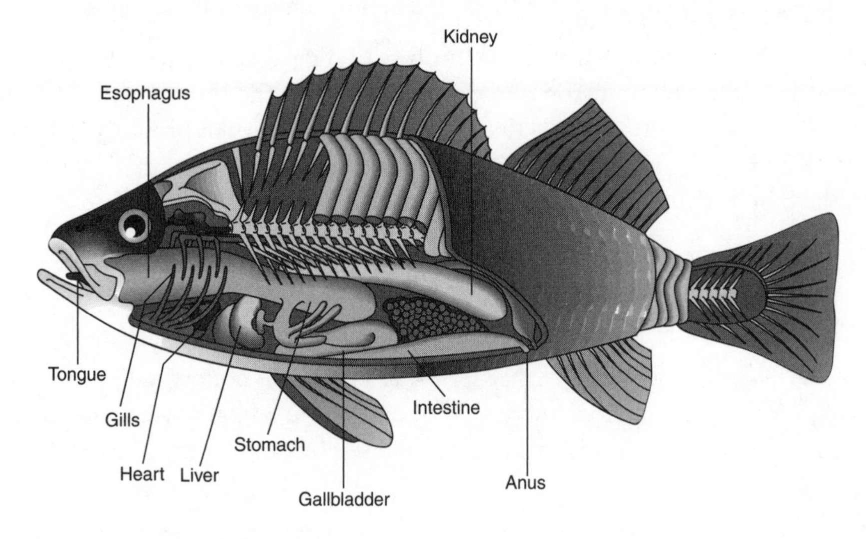

Fish digestive system.

Amphibian Digestive System

Amphibians have a herbivore juvenile period and a carnivore adult period, which necessitates a complex digestive system. A herbivore feeds on vegetative materials, while a carnivore feeds on animal food sources.

Tadpoles are the larval stage of frogs and toads. They filter-feed small amounts of algae and organic matter from the aquatic environment. Because tadpoles must grow rapidly to capitalize on spring puddles and algae blooms, they often stuff their digestive system with hard-to-digest vegetative matter. To compensate, their intestinal length is increased to accommodate the extra time needed for digestion.

Frogs and toads are typical amphibians. Their carnivorous adult forms and their digestive anatomy is similar to more advanced animals. Their system includes the following, in order of function:

- Mouth, pharynx, and esophagus for catching and transferring food.
- Stomach, which receives a large amount of food from the elastic esophagus and stores the food while stomach cells secrete digestive enzymes to begin the chemical digestion of food.

- A circular, smooth muscle called the *pyloric sphincter* fits around the digestive tube between the stomach and intestine. When constricted, it holds the contents of the stomach in place for digestive purposes. When the digestive action is complete, it relaxes and the slurry slides into the small intestine.
- The *small intestine* continues the process of digestion and accepts *bile*, which is made in the *liver* and stored in the *gallbladder* for the breakdown of fats and a host of pancreatic enzymes that digest different types of proteins and carbohydrates. Digested nutrients are absorbed by specialized cells that transfer it to the bloodstream. The remainder is pushed into the large intestine.
- The *large intestine* reabsorbs any excess water and collects the indigestible waste products. Smooth-muscle activity rhythmatically moves the now-solid mass into the final internal location.
- The *cloaca* is a body cavity that serves as a common collecting point for solid digestive waste, liquid urinary wastes, and sperm and egg from the gonads. From the cloaca, the substances are released from the body.

Bionote

Sphincter muscles also exist in humans. For instance, the pyloric sphincter regulates the flow of materials out of the stomach. The *cardiac sphincter* regulates food flow between the esophagus and stomach, the *ileocal sphincter* operates between the small and large intestine, and the *anal sphincter* regulates the flow of materials out of the anus. Most sphincters are controlled by the autonomic system; however, in humans and some other mammals, the anal sphincter is actually two separate sphincters. The internal sphincter is involuntarily controlled, while the external sphincter is under voluntary control. The control of the external anal sphincter is a learned behavior, often called "potty training" (see Chapter 19).

Except for the cloaca, the frog digestive system is similar to the humans. In addition, the digestive systems of birds and reptiles are not significantly different from those of fish and frogs.

Mammalian Digestive Systems

Mammals have the most efficient digestive system. They are among the first animals to chew their food, which mechanically breaks down the food while digestive enzymes in the mouth begin the chemical digestion. Chewing speeds up the process because it creates more surface area on the food for the enzymes to act. Whales are an exception; they strain ocean water through thin, fingerlike projections from the roof of their mouth called *baleen*. Any organism trapped in the baleen filter is swallowed.

Vegetative materials are hard to digest for mammals because of the cellulose. In fact, no vertebrate contains an enzyme to digest cellulose! So why do vertebrates bother to eat plants? Interestingly, a symbiotic relationship exists between certain microorganisms and vertebrates. Certain bacteria are able to break down cellulose and live symbiotically in the gut of most vertebrates. Most hoofed mammals have multiple stomachs that serve a variety of functions, including the digestion of plants. Usually the first stomach, or *rumen*, contains the microorganism. Cows are well known for regurgitating food back and forth from the rumen to the mouth to rechew the food to increase the surface area for increased microbial action. When a cow chews its cud, it is rechewing partially digested food. Horses, elephants, and rodents have a *cecum*, which is a saclike structure hanging off of the end of the small intestine. They use the cecum for digestion of vegetative matter in the same way. Humans also have a cecum and a saclike structure extending from the closed end of the cecum called an appendix; it's usefulness as a digestive structure is questionable.

Human Digestive System

The human gastrointestinal tract is not much different from that of lower animal forms. The major noticeable difference is the number of and specificity of certain enzymes allowing humans a wide range of food sources.

In the mouth, a combination of teeth types and saliva begin the process. Humans possess both incisors for cutting, molars for grinding, and a muscular tongue to keep the food in proper chewing alignment. Saliva is the watery solution that adds mucus and the digestive enzyme salivary amylase to the food. The mucus lubricates the food mass for transfer to the next processing station.

The rhythmic action of circular and longitudinal muscles called peristalsis pushes the food down the esophagus. The *epiglottis* is a flap of cartilaginous tissue that directs air into the lungs and food into the stomach. When people choke on food, it is usually because they tried to talk or laugh and swallow at the same time, thereby keeping the trachea open and allowing food to enter the windpipe.

The membrane of the stomach is lined with gastric pits that contain specialized tissue that secrete digestive enzymes, mucus, and hydrochloric acid into the stomach.

The acidic gastric fluid converts pepsinogen into the protein-digesting enzyme pepsin, which breaks polypeptide chains into shorter chains. Interestingly, the mucus secreted by the stomach is alkaline, which protects it from its own harmful stomach acids.

The liver delivers bile to the small intestine via the gallbladder, and the pancreas delivers digestive enzymes to the small intestine similar to the frog's mechanism. In humans, the small intestine is often 21 feet (7 meters) in length. Mucus secretions in the small intestine protect the cells from the action of the enzymes and the alkaline nature of the mixture. Digestion of proteins, fats, and carbohydrates is completed in the small intestine. Nutrients are continually absorbed in all parts of the small intestine. Additionally, almost all water is reabsorbed in the small intestine. The liquid *chyme* is then moved downward.

Bionote

An ulcer is a lesion created when the stomach acid penetrates the protective mucus and begins destroying the stomach lining.

The large intestine functions mostly to reabsorb water and any remaining nutrients. The large intestine also absorbs the products of microbial digestion that took place in previous organs as well as the microbial products produced in the large intestine, which is the major site of microbial digestion. As water is absorbed from the chyme, it becomes more solid. As more material is added to the mass, it causes a distension of the intestine, which begins the muscular contractions that drive the fecal materials to the rectum. Mucus is constantly added to lubricate the passage from the rectum to the anal opening into the environment. The digestive system takes raw food material and systematically breaks it down into usable nutrients. Digestion often includes both mechanical and chemical processes. The length of the system indicates the sophistication of the specialization.

Invertebrate Excretory Systems

Even though the poisonous nitrogen compound ammonia is produced as a waste product of metabolism in invertebrates as well as vertebrates, the invertebrates have a simplified system for solving the problem. Because ammonia is very soluble in water, aquatic invertebrates, including porifera, cnidarians, and nematodes, simply diffuse the ammonia into the surrounding water. Most soft-body invertebrates, such as flatworms, diffuse ammonia throughout their entire exposed surface. This rate of ammonia disposal is fast enough to prevent the toxic ammonia from building to a lethal concentration in small invertebrates.

Bionote

In most cases, the excretory system is used to eliminate nitrogen compounds from the system; however, in several species of freshwater flatworms, specialized cells called flame cells are designed to remove excess water from the body that enters via osmosis.

Nephridia and Malpighian Tubules

Nephridia and malpighian tubules are specialized structures designed to remove nitrogen compounds from the system. In both structures, the toxic ammonia is converted to a less-toxic form. For terrestrial animals, removal of ammonia into the air does not occur fast enough to prevent death. Annelids, molluscs, and invertebrate chordates employ nephridia. The tube-shaped nephridia receive body fluids and osmotically return most of the water to the body. The nitrogen compounds are collected, converted to the less-toxic compound urea, and excreted in the urine from the body.

Malpighian tubules are characteristic of arthropods and are similar in function to nephridia. Malpighian tubules are fingerlike projections that extend from the arthropod's gut. They are strategically located to receive a continual supply of fresh blood from which they extract water and dissolved wastes, such as ammonia. As the solution is pushed through the tubules, most of the biologically important ions, water, and nutrients are reabsorbed by osmotic pressure back into the body. The malpighian tubules then convert the dangerous nitrogen compounds into the less-toxic compound uric acid, which can be stored in the digestive tract until deposited through the anus into the environment. Both nephridia and malpighian tubules separate nitrogen compounds for removal from the blood. In the process of creating waste urine, water balance is retained along with important ions and nutrients.

Water Retention

Aquatic animals—such as marine invertebrates, whose body-solute concentration is the same as their environment—are *osmoconformers* and never worry about water balance because water moves osmotically to equalize concentrations. For all other animals, *osmoregulators*, water balance is a continual problem. Fish have a slight advantage, however, because they live in water. Freshwater fish have to get rid of excess water because their internal solute concentration is greater than the surrounding water, which creates an osmotic gradient that draws in water through their gills. To solve this problem, they do not drink water and they discharge diluted urine. Saltwater fish have the opposite osmoregulatory problem. To maintain homeostasis, they drink water and diffuse salt out via gills. They also concentrate their urine to eliminate more salt per unit of water.

Vertebrate Excretory Systems

Land animals struggle against dehydration in a number of ways. Primarily, they (we) drink water as needed. Some arthropods have waterproof exoskeletons to eliminate evaporation loss; others avoid direct sunlight or heat; others create watertight protective containers for eggs and embryos.

Vertebrates continue the progression of increasing complexity and efficiency in their creation of organs and organ systems that accommodate removal of nitrogen and the retention of water. These advancements have allowed vertebrates to colonize fringe areas that greatly extend their territories. The most complex vertebrate, man, demonstrates the summation of organs and organ systems developed in simpler animals. Regardless of the complexity of the function, all systems operate on fundamental scientific principles.

Fish Excretory System

Fish have *kidneys* that function to filter ammonia and other dissolved wastes from the blood. In a sense, they are filters that clean the blood. By filtering the blood, they help regulate the concentration of various ions in the system, which helps maintain homeostasis for the entire body. The functional unit of the kidney is the nephron. Refer to the illustration *A typical nephron* in the "Nephron Structure and Function" section.

Fish also diffuse ammonia directly back into the water through their gills. For most fish, diffusion by gills has to be supplemented with nitrogen removal by the nephrons. The nephrons also function to maintain the salt and water balance for the fish. Essentially, the nephrons accept all blood flow and return water, dissolved biologically important ions, and cellular substances back into the bloodstream while removing harmful wastes. The nephrons remove the ammonia and other filtered wastes in a urine solution. The water and salt balance is regulated by the reabsorption of the water and salts in the nephron.

Human Excretory System

The metabolism of proteins releases toxic nitrogen compounds in mammals and humans as it does in the remainder of the animal world. Removal of dissolved nitrogen compounds without the loss of useful salts and water is a problem for all animals. Both mammals and man have employed structures and techniques first utilized in much simpler animals.

Mammals convert nitrogen compounds into urea in the liver. This process requires energy, but it creates a less-toxic form of nitrogen that can be stored longer without harming the organism.

In humans, dissolved ammonia in the bloodstream is captured by the liver and converted into the less-toxic nitrogen compound urea and then is released back into the blood. The urea is then filtered from the blood by the kidneys. The kidney function is so important that everyone has two located on both sides of the spine in the lower back. The hand-sized, kidney-bean-shaped structures also have a renal artery bringing blood directly to both of them. The kidneys are responsible for maintaining salt and water balance while eliminating poisonous wastes.

Nephron Structure and Function

The main function of the human kidney is to maintain water balance and filter unwanted substances out of the blood and into the urine for discharge. The active site in the kidney is the *nephron* and also represents a classic case where structure fits function. The primary functioning structures of the nephron are, in order, glomerulus, proximal tubule, Loop of Henle, distal tubule, and collecting duct. They perform the keys functions of filtration, reabsorption, secretion, and excretion. Refer to the following illustration *Typical nephron* as you think about this next section.

The following are the five steps describing the nephron function:

1. Blood pressure pushes water solutes and urea, but not protein, out of the blood contained in the ball of capillaries called the glomerulus into the nephron, creating the initial filtrate. The remainder of the blood continues to circulate.
2. The proximal tubule is primarily designed for the recovery of water. However, most of the salt, glucose, and amino acids are also recovered and returned to the bloodstream. Waste compounds remain in the tubules.
3. The entire nephron is contained within the renal cortex and deeper renal medulla, two adjoining structures that vary in their solute concentration. The Loop of Henle connects with the proximal tubule in the renal cortex and then makes a hairpin loop in the renal medulla and connects with the distal tubule in the renal cortex. As the filtrate travels down the loop, the concentration of solutes in the renal medulla increases, thereby drawing water osmotically out of the filtrate. The water is simultaneously removed by the circulatory system, thereby maintaining the osmotic gradient. After the hairpin Loop, the tubule becomes impermeable to water, so only dissolved solutes can diffuse out and enter the bloodstream.
4. The *distal tubule* reabsorbs needed salts, bicarbonate, and other nutrients as required by the body. Both proximal and distal tubules can secrete hydrogen ions and certain poisons, like ammonia, into the filtrate; however, only the distal tubules can secrete medicinal drugs into the filtrate.

Bionote

Dialysis is a technique to machine-filter blood for people with kidney problems. Dialysis works in a similar fashion as the nephrons in the kidney by diffusing out unwanted blood components via a circulating dialysis solution and returning desirable substances in the same manner.

5. All distal tubules empty into a collecting duct that further removes any salts, water, or urea from the filtrate. The collecting duct is the only structure capable of recovering urea as needed, especially in the renal medulla to maintain the osmotic gradient to facilitate the removal of water from the filtrate. The filtrate still retains some of the nitrogen wastes filtered earlier.

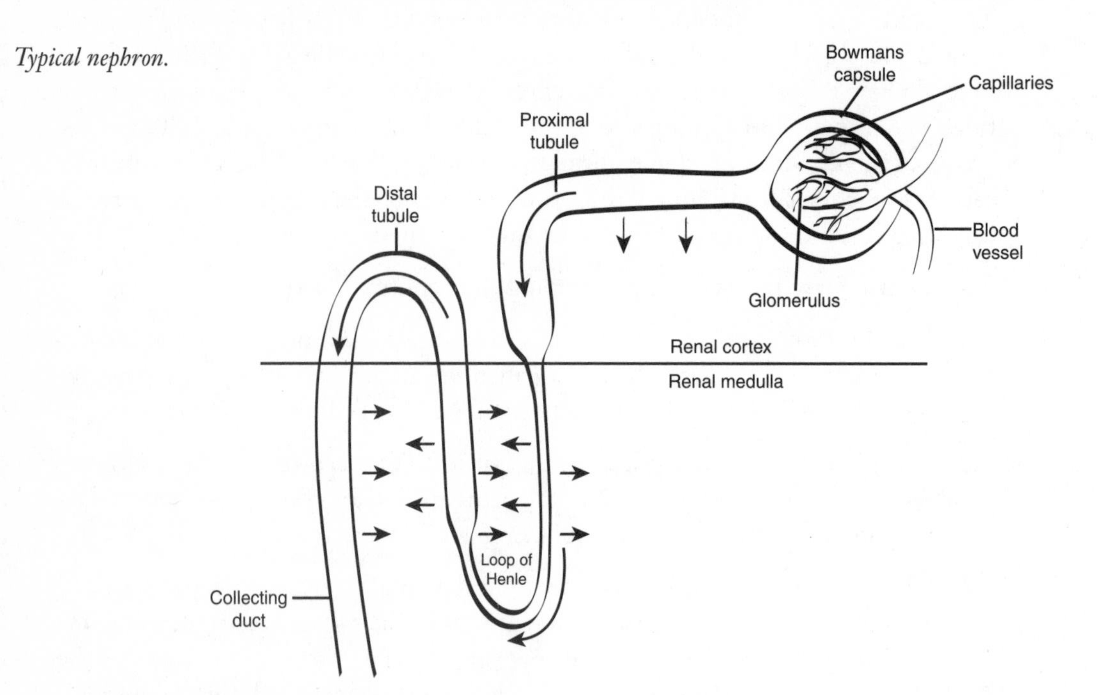

Typical nephron.

The human digestive and excretory systems are interesting studies on the increasing complexity of the animal hierarchy. In both cases, simple processes are included into bigger and more sophisticated structures that accomplish the same task but on a larger scale.

The Least You Need to Know

- Unicellular organisms absorb food and digest it internally.
- A gastrovascular cavity allows cell specialization and digestion outside of the cells.
- The alimentary canal creates areas of digestive specialization.
- Organs contribute digestive enzymes throughout the process to increase digestive opportunities.
- The human digestive system is not much different from that of other advanced animals.

Chapter 15

Endocrine, Immune, and Nervous Systems

In This Chapter

- How the endocrine system regulates body functions
- The hypothalamus is the "master gland"
- The nervous system of simple invertebrates
- How neurons transmit a nerve signal
- Why some people seem to catch disease easier than others
- The role of the circulatory system in the immune system

Continuing the study of how your body works, we move on to the endocrine, nervous, and immune systems, which help maintain your sense of well-being. They do this by adjusting internal biological reactions, and fighting invasive agents before they have a chance to give you an unwanted day off from work, or play.

The endocrine system contains the ductless glands that secrete hormones that run rampant in teenagers. There are several clever mechanisms that describe how the various hormones work inside you.

The nervous system is the electrical wiring for your body. It is responsible for picking up important biological and other stimuli from the external and internal environment and transferring it to some organ that can monitor the data and react, if necessary. What would we do without it?

The immune system is the health-safeguard system that keeps you going even when under attack from germs and worms. Your body has three lines of defense that invaders must defeat before you begin to feel their effects.

Endocrine System Functions

The *endocrine system* in vertebrates is a series of ductless glands that secrete hormones. *Hormones* are chemical compounds that are messengers released from endocrine glands that usually travel in the bloodstream to effect an action somewhere else in the body. Although they are sometimes released into the interstitial fluid of neighboring cells for their function, typically hormones travel at the rate of blood flow to prompt an enzyme into action.

The activity of the endocrine system is exhibited by the functions of the two major types of hormones:

- *Peptide hormones*, also known as amino acid-based hormones, are water soluble and are made from modified versions of single amino acids, which are called *amino hormones*. Peptide hormones are made from small segments of an amino acid chain; the largest are the *protein hormones* that are made from long chain polypeptides.
- Steroid hormones are fat-soluble, lipid hormones made in the body from cholesterol.

Both peptide and steroid hormones normally travel in the bloodstream and chemically attach to specific *receptor proteins* in *target cells*, which subsequently change shape to perform their function. Their mechanism is slightly different.

Peptide Hormonal Control

Peptide hormones are water soluble and are generally quite large, which prevents them from passively entering the cell, so their mechanism is external. Refer to the illustration *Peptide hormone mechanism*.

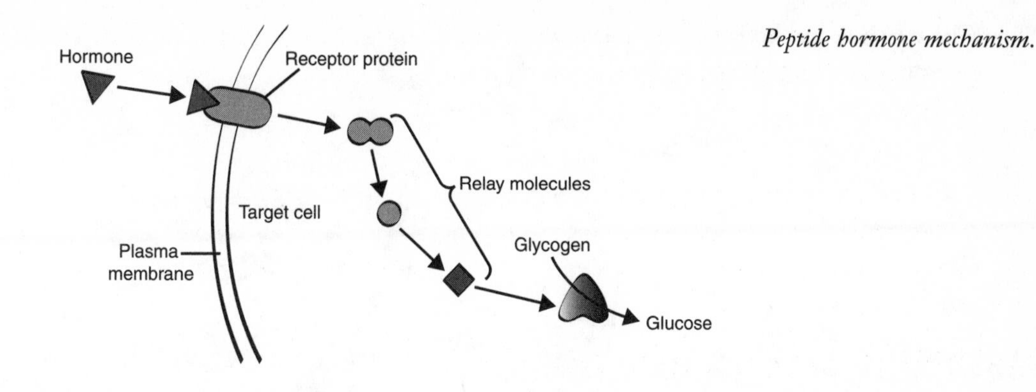

Peptide hormone mechanism.

The mechanism of a typical peptide hormone is described in the following steps:

1. The hormone binds to a specific receptor protein in the plasma membrane of the target cell that changes the chemical structure and shape of the receptor protein, which activates it.

2. The activated receptor protein in turn activates an enzyme in the membrane, which creates a "second messenger." The second messenger usually catalyzes a series of reactions that terminates with the activation of the target cell.

The common example of peptide hormone action is the effect of glucagons on glucose production. Glucagon is made in the pancreas and binds with the appropriate receptor proteins in liver target cells, which in turn alters their shape. This structural change releases a membrane enzyme that converts the available cellular ATP into the second messenger, cyclic AMP, which then initiates a series of cytoplasmic, enzyme-controlled reactions that convert glycogen into glucose. Typical of peptide hormone mechanisms, the hormone causes a cellular event without even entering the cell!

Steroid Hormonal Control

Steroid hormones are fat soluble and therefore easily penetrate the phospholipid cell membrane of their target cells, so their mechanism is internal. Refer to the illustration *Steroid hormone mechanism.*

The mechanism for a typical steroid hormone is described in the following steps:

1. The hormone binds to a specific receptor protein in the cytoplasm to create a hormone-receptor complex, which enters the nucleus and attaches to specific sites on the DNA.

2. Binding the hormone-receptor complex to the DNA either activates the transcription of certain genes into RNA or prevents the process of transcription and ends the process.

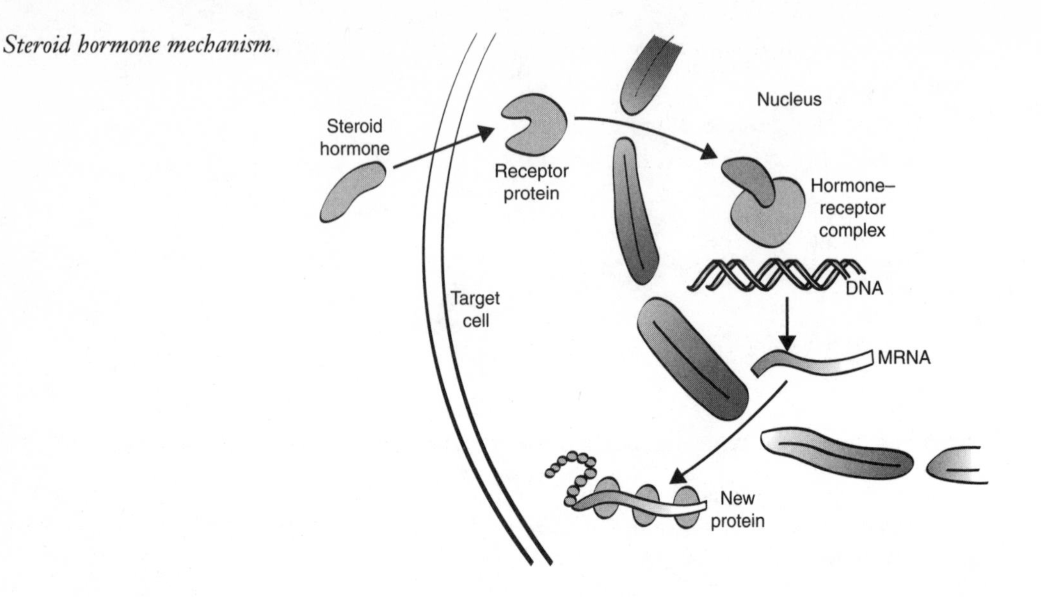

Steroid hormone mechanism.

Steroid hormones regulate gene activity, and unlike peptide hormones, always produce a genetic event. The most common example of the steroid hormone mechanism is the gene regulation produced by the sex hormones. Estrogen is made in the ovaries and testosterone in the testes, both of which are able to diffuse through a target cell membrane and attach to their respective receptor protein site. The resulting complex binds to the appropriate DNA site, which activates the transcription of the corresponding genes into RNA. The transcribed RNA translates the message into the proteins responsible for either the male or female secondary sex characteristics!

Bionote

Only the sex cells and adrenal cortex gland produce steroid hormones.

Typical Hormone Mechanism

The *hypothalamus* is sometimes called the central hormone center because it receives electrical impulse information from the nervous system and chemical information from blood hormones and then directs a response in the form of a chemical messenger to the pituitary gland. The hypothalamus is located in the center of the skull for extra protection, and at the base of the brain. The pituitary gland appears to be an extension dangling beneath and connected to the hypothalamus. The pituitary gland produces nine hormones that are directed by six hormones from the hypothalamus. The six hypothalamus hormones direct the anterior area of the pituitary gland to release or stop releasing pituitary hormones. The hypothalamus is the receiving-processing center that hormonally dictates a response to the pituitary gland. Neurosecretory cells in the hypothalamus produce hormones that are transported to the pituitary gland where they effect the action of the gland. The pituitary gland then sends a hormone response to the area of concern.

Bionote

The hypothalamus also has a nervous system connection to the posterior area of the pituitary, which releases vasopressin, the antidiuretic hormone.

The hormones of the pituitary gland affect and interact with most other glands, especially those responsible for growth.

A common example of a hormone triggering a subsequent body activity is the action of the pancreatic hormones, insulin and glucagon, on the availability of glucose in the bloodstream for intake by cells. Digested foods, including glucose, elevate the blood sugar level as they enter the bloodstream. The rise in blood sugar level triggers the action of insulin, a peptide hormone. Insulin activates a receptor protein that transports glucose in the blood through the cell membrane into the cytoplasm. Insulin works because it takes sugar molecules from the bloodstream and enables them to pass through the cell membrane. By taking sugar out of the blood, it lowers the blood sugar level and provides glucose for cellular respiration. Glucagon works antagonistically with insulin so that a starving person releases glucagon molecules from the pancreas, which activates the conversion of glycogen stored in the liver to glucose. The conversion of glycogen to glucose and the release of the newly constructed glucose molecules into the blood raises the blood sugar level.

Diabetes mellitus Type I is a hereditary disease that destroys the insulin-producing section of the pancreas. The resulting insulin deficiency raises the blood sugar to dangerous levels. Approximately 10 percent of the North American population is insulin dependent, meaning that they have to supplement their normal insulin levels with daily insulin injection. Failure to sustain proper insulin levels forces the kidneys to excrete glucose (sugar in the urine is often a test for diabetes), which is accompanied by the osmotic loss of water, creating a continual thirst. The body is also forced to degrade proteins and fats for energy, which creates acidic compounds that build up in the blood. The resulting increase in blood acidity leads to a diabetic coma and sometimes death.

The endocrine system is integrally linked with the nervous system to maintain homeostasis. So why are two systems needed? The response time for most endocrine hormones is dependent on blood flow, making it slower than nervous transactions. Likewise, nerve-generated hormonal actions tend to be short-lived, whereas most endocrine hormone actions are more long term.

Sensory and Nervous Systems

Within the animal kingdom, the nervous system functions basically the same: reception of internal and external stimuli by receptors; transmission of the message to a processing center; integration of data; response from the processing center to parts of the body for action.

Evolution of Nervous System

Although they may function similarly, the degree of sophistication varies with the complexity of the animal. For instance, Porifera are so simple that they do not require, or have, a nervous system, whereas cnidarians such as the common hydra have a network of neurons called a *nerve net* throughout their body. Likewise, echinoderms are even more advanced, with clusters of neurons acting in concert to form *nerves.* Cnidarians and echinoderms are radially symmetric, which orients their nervous system accordingly, with no definite head or centralized processing center.

Flatworms are the simplest bilaterally symmetrical animal to exhibit *cephalization* and *centralization* of their nervous system. Cephalization is the concentration of the nervous system in a central processing and coordinating area, or head. Centralization is the possession of a central nervous system (CNS), which is structurally separate and functionally integrated to the peripheral nervous system (PNS). The CNS includes the brain, and in vertebrates, the spinal cord, whereas the PNS includes the remainder of the nervous system. Increasingly, cephalization and centralization indicate greater animal complexity.

Diversity of Nervous Systems

Increased diversity in sensory receptors and sophisticated processing of their data is also a sign of increasing complexity. There are five main types of sensory receptors:

- A *chemoreceptor* is usually located in the nose and tongue and distinguishes chemicals, such as between foods and nonfoods, whereas other chemoreceptors located in other body parts may signal environmental changes such as fire.
- *Mechanoreceptors* register movement pressure and changes in skin tension. Located throughout the skin, they are concentrated in high-contact areas such as the face and fingertips.
- *Pain receptors* are a survival advantage because they warn the body of harmful conditions causing possible tissue damage. They often work in concert with other receptors to create reflex actions to painful stimuli such as touching a hot object.

- *Photoreceptors* respond to light and are usually found in the cephalic region of animals, such as the simple protist euglena.
- *Thermoreceptors* detect temperature changes and are located everywhere in the skin. They are also a classic example of nervous system/endocrine system co-operation as they both coordinate through the hypothalamus.

Neuron Structure and Function

No matter which sensory receptor is stimulated, or in what type of animal, the sensation is transmitted in the same mechanism as described next. Refer to the illustration *Neuron structure and function.*

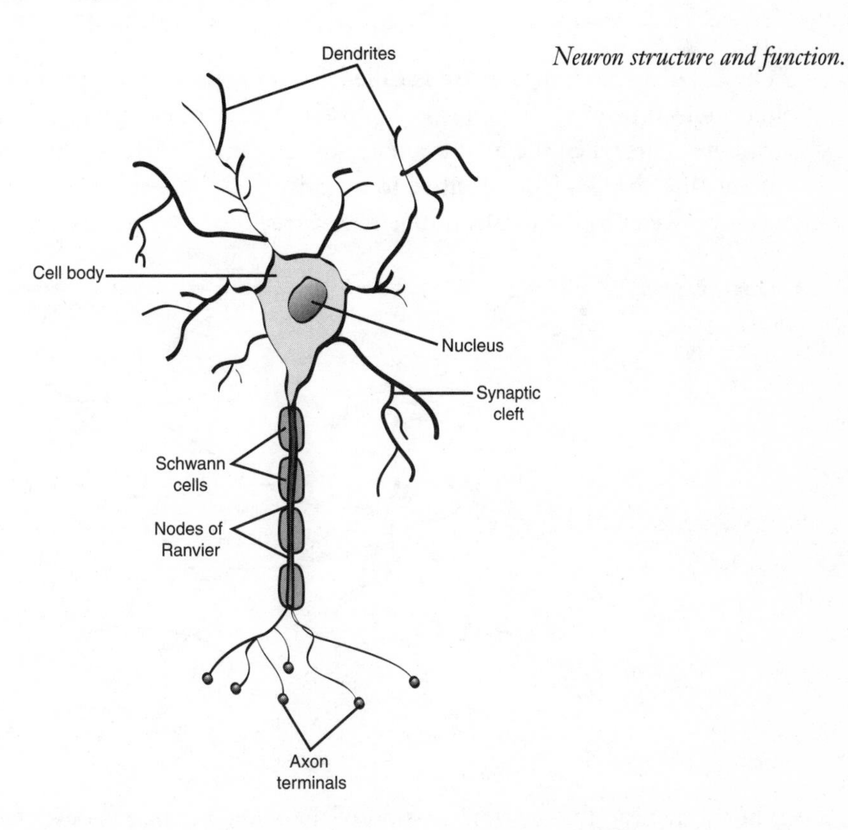

Neuron structure and function.

The active site of all nervous systems is the neuron. *Sensory neurons* send signals from sensory receptors to the CNS; *interneurons,* located within the CNS, help to aggregate and interpret the signals and then to relay them to other interneurons or *motor neurons,* which transmit the signals to the response area. So how is the signal transmitted?

The neuron is structured solely to receive and transmit signals. A wave of depolarization proceeds from its receptive end, called the dendrite, to the receiving end, called the axon. It occurs because a normal neuron has a *resting potential* across its cell membrane of 70 millivolts because of the charge difference between the positive ions outside and negatively charged ions inside the cell. This charge overload and separation creates a charge differential across the membrane.

Bionote

To show the size contrast, D-cell batteries have a potential of 1.5 volts; a millivolt is 0.001 volts.

The membrane of the neuron remains at its resting potential unless interrupted by a stimulus. During the resting period, an ATP-assisted sodium and potassium pump actively moves positively charged sodium ions (Na^+) out of the cell and positively charged ions (K^+) into the cell to maintain the polar nature of the membrane and its resting potential.

The actual signal transmission is called a *nerve impulse* and moves in one direction like a one-time wave. The nerve impulse is a wave of charge depolarization. Depolarization occurs when the membrane changes charges, such as when the outside of the membrane changes from positive to negative and the inside changes from negative to positive. Refer to the illustration *Mechanism of impulse transmission.*

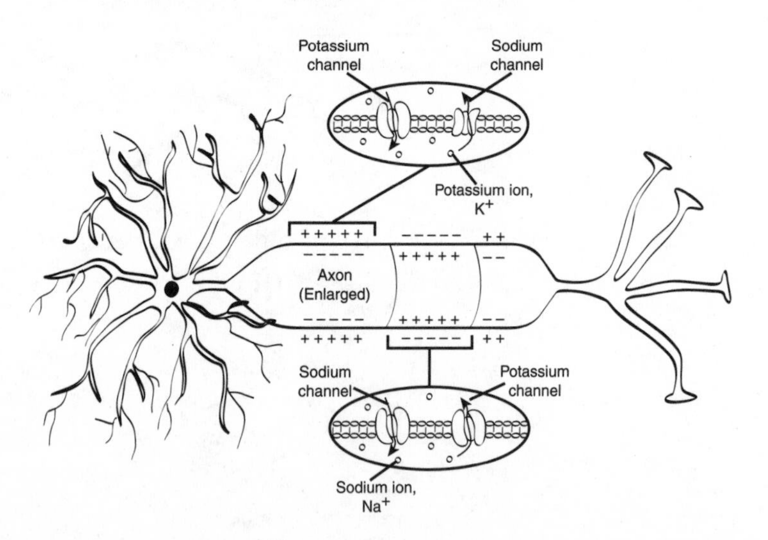

Mechanism of impulse transmission.

The mechanism for the transmission of the nerve impulse is described in the following steps:

1. At resting potential, the neuron membrane separates the positive charge outside the cell from the negative charge inside the cell until stimulated by the leading edge of a nerve impulse. The nerve impulse opens specialized protein channels

in the cell membrane called *sodium gates* that allow Na^+ to move quickly into the cell, thereby depolarizing the cell membrane.

2. If the impulse is greater then –50 millivolts, the *threshold potential,* it triggers adjoining Na^+ gates to open, thus creating the unidirectional flow from dendrite to axon. If the potential is less than –50 millivolts, the impulse is not transmitted.
3. When the impulse passes, K^+ gates open allowing the K^+ to move outside the cell, thereby repolarizing the cell membrane; positive outside, negative inside.

The depolarization and repolarization happen very quickly, allowing for another wave of depolarization, also called an *action potential.* A wave of depolarization further defines a nerve impulse as a wave which propagates an action potential moving along the membrane. The action potential has been likened to a row of dominoes arranged so that knocking over the first domino pushes it into the second domino until all are knocked down. The only difference in this metaphor is the repolarization happens faster than the dominos can be set up again.

Note that the axon in one neuron does not touch the dendrite of another; instead, a gap or *synapse* separates them. The impulse moves across the synapse in one of two ways, either electrical or chemical.

In chemical synapse crossings, the electrical impulses are converted into chemical signals called neurotransmitters, which are specialized molecules stored in vesicles at the terminal end of an axon. The action potential stimulates the release of the neurotransmitters, such as acetylcholine, to diffuse across the synapse and bind to protein receptor molecules on the dendrite of the next neuron. The binding opens the Na^+ gates, which regenerates the action potential in the receiving cell. The neurotransmitters are then either reabsorbed by the axon or degraded by specific enzymes.

Electrical synapse crossings are much easier and much faster and are therefore used to conduct emergency signals, such as fight of flight. In this case, the actual potential moves from neuron to neuron across a synapse. Electrical synapses are the exception, not the norm. Another structural advancement in vertebrates is the support cells collectively called *Schwann cells* that surround the neurons for insulation. Schwann cells also link together to form *myelin sheaths,* which are accumulated Schwann cells that wrap around the neurons. The myelin sheath also speeds up the transmission of the action potential by allowing it to jump from spaces between adjoining sheaths called nodes of Ranvier. The signal can jump from node to node faster than normal transmission because it is not slowed down by having to internally travel the full length of the axon. The rate of transmission with myelin is 150 m/sec, or 330 mph, whereas without myelin the rate is reduced to about 5m/sec or 11 mph. Multiple sclerosis in humans is a disease that destroys large areas of the myelin sheath around the neurons in the brain and spinal cord, thereby impairing their delivery of impulses.

Immune System

There are three methods by which a body protects itself from disease; two are generic, whereas the third is pathogen specific.

Multicombination Mechanisms

The first line of nonspecific defense against *pathogens* (anything that causes a disease) consists of eight separate events:

1. Skin is impermeable to water and most other types of potential intruders.
2. Sweat, saliva, and tears contain the enzyme lysozyme, which destroys the cell walls of many bacteria. Additionally, oils secreted by sebaceous skin glands and the acidic nature of sweat prohibit multiplication by microorganisms.
3. Mucous membranes line the two organ systems that connect with the environment, digestive and respiratory, and produce a viscous liquid called mucous that traps uninvited particles.
4. The acidic liquids found in the stomach juices kill almost all of the bacteria and protists that are ingested with food.
5. Nose hairs and respiratory cilia trap foreign particles and often hold them until expelled or engulfed by mucous.
6. Whereas blood cells are probably the most common, they are an amalgamation of many different types of cells that circulate in the blood, lymph, and interstitial fluid. For in stance, *neutrophils* surround and destroy invaders, whereas *macrophages* are larger white blood cells that engulf larger particles in the interstitial fluid.
7. Virus-infected cells produce a specialized protein called *interferon* that is released to signal defense systems in neighboring cells.
8. *Complement proteins* circulate in the bloodstream and become activated by invaders or the immune system to destroy the invaders or alter the protein coat of the invader for easier identification by neutrophils and macrophages.

Bionote

Recombinant DNA has provided mass production of interferon, which is used to treat certain viral infections and as a possible cancer-inhibiting agent.

Inflammatory Mechanism

The nonspecific inflammatory response has a three-step mechanism:

1. Histamine is released by most injured cells, causing the blood vessels to dilate. Blood flow increases, and the blood vessels become leakier as they dilate.
2. Blood plasma containing white blood cells leaks into the interstitial fluid. (Pus is mainly dead white blood cells and interstitial fluid that occurs if there is a bacterial infection.)
3. The white blood cells destroy the invaders and they engulf and remove any damaged cell parts.

Sometimes if the disease is widespread in the body, the inflammatory mechanism institutes a fever, which may be triggered by the toxins themselves or by certain white blood cells and is designed to activate the white blood cells. As a result, the white blood cell count is often used by medical practitioners to diagnose a disease.

Immune Mechanisms

The immune system is a specific defense activated by the presence of a specific *antigen* and retains a memory of that antigen for an indefinite period of time, even after the infection is cured.

There are two types of *lymphocytes*, which are white blood cells: B-lymphocytes (B cells) and t-lymphocytes (T cells), which cooperate to form the active part of the immune system. B cells secrete *antibodies* that fight against invaders in the interstitial fluid, and T cells attack any infected body cells.

The B cell mechanism involves two steps:

1. The B cell binds with the antigen, triggering the production of more B cells.
2. Two types of B cells are produced: plasma cells, which secrete antibodies as the first response; and memory B cells, which retain the shape of the original B cell and are considered the second response to an invasion because they are able to resume production of both plasma cells and new memory B cells after long periods of inactivity after the disease is destroyed.

Bioterms

An **antigen** is a molecule or system of molecules usually located on the exposed surfaces of cells, such as viruses, bacteria, cancer cells, and pollen. An **antibody** is a blood plasma protein that binds with a receptor site on an antigen and works to destroy it.

One type of T cell responds to antigens attached to the surface of existing body cells. There are two types of T cells that are of interest in this mechanism: helper T and cytotoxic T cells.

Helper T cells have a fascinating mechanism that involves interaction with a specialized macrophage called an antigen-presenting cell (APC) because of its function. The APC engulfs an invader and degrades it into smaller units. Then specialized proteins of the APC bind with the fragments of the invader and present them on the cell surface to attract helper T cells. Apparently, the helper T cells only bind when both the APC protein and invader fragment are displayed. The binding of the helper T cell activates it to produce proteins such as interleukins that stimulate the growth and division of the helper T cells. When this happens, the helper T cells then produce more helper T cells and B cells, and activate the cytotoxic T cells.

Cytotoxic T cells identify infected cells with the same double identification system employed by helper T cells: An infected cell has its own proteins on its surface combined with the invader's proteins. The cytotoxic T cell is attracted to and combines with this complex. Binding to this complex activates the cytotoxic T cell to release perforin, a specialized protein that drills a hole in the infected cell's membrane allowing the contents to spill out and cause cell death.

The endocrine, immune, and nervous systems are an interesting study of interrelated processes. The overriding principle is the health of the organism. All of the systems combine and coordinate their efforts to maximize their value.

The Least You Need to Know

- The immune system for humans has three separate mechanisms for defensive action, one of which is invader specific.
- The circulatory system and the immune system provide an active defense against infections that encompasses the whole body.
- Endocrine glands are ductless, but they secrete hormones that travel via the bloodstream to effect an action somewhere else in the body.
- The hypothalamus receives both chemical and electrical input, which triggers endocrine reactions.
- The nervous system has evolved to provide better sensory input from the environment and quicker responding actions as animals became more complex.

Chapter 16

Skeletal and Muscular Systems

In This Chapter

- The function of a skeleton
- Skeleton on the inside or on the outside
- Cartilage skeleton versus bone skeleton
- Different muscle types
- Muscle tone and how it is improved

You have 206 bones in your body; so does a giraffe. They function in the same way: to hold you in place and provide a place for muscles to attach. Invertebrates do not have a backbone, nor can they stand up. A giraffe is able to move because the skeletal muscles pull on the bones until they move; your system works the same way.

A very detailed analysis of how muscles actually work is presented with pictures to help. The mechanics are quite clever. The muscular system is made up of specialized muscle tissue that functions to contract when stimulated by sliding past one another. The shortening of the muscle is a contraction and that is what moves a bone, or pumps the heart, or moves food products through the digestive system. It can be easy when you examine it in stages, as we do in this chapter.

Skeletal System

These are three types of skeletal systems employed by animals. Each type reflects the general purpose for that organism, but in general all skeletal systems provide structure and support for the body as well as protection of internal organs. A *hydrostatic skeleton* is in cnidarians and certain worms such as the common earthworm. A hydrostatic skeleton contains no bone or hard structures, and functions under the pressure of fluid in a closed system. Because of the need for increased surface area for gaseous exchange and food collection, a hydrostatic skeleton cannot support upright terrestrial lifestyles.

An exoskeleton is the trademark of arthropods such as insects. The external skeleton provides overall protection and support. Muscles are attached to the interior of the exoskeleton, allowing a range of movements. In arthropods, specialized cells secrete the exoskeleton, but the shell itself is nonliving. Because of this, as the animal grows, it has to shed its old shell and grow a new one in a process called *molting*. One of the most vulnerable times for a molting animal occurs after it leaves the old shell and before the new shell is hard enough for protection.

Bionote

Watermen harvest the Maryland blue crab during the warm summer months as they begin to molt for a local delicacy called soft crabs.

Vertebrate *endoskeletons* are sometimes made of cartilage (sharks, for instance) or cartilage-bone combinations (humans, for instance). The human skeleton is designed to support an upright terrestrial stance and is one of the reasons why man is able to walk on two legs and proceed in relative safety from most normal accidents.

The human skeleton consists of 206 bones and associated cartilage organized into two types of skeleton:

- The *axial skeleton* is common to all vertebrates and includes the skull, backbone, and rib cage. The axial skeleton's purpose is primarily the protection of soft internal organ tissue.
- The *appendicular skeleton* provides support for the body and structures for its functions. The pelvic girdle and legs are vertically oriented to support an upright stance. The shoulder girdle and the arms are designed for work.

Bones

In humans, actual bones begin to form when the fetus is 6 weeks old and continue to grow until the DNA-specified height is reached (at 18 years for females and 21 years for males). Bones are living tissues that require adequate nutrition and waste removal, so they are well supplied with blood vessels. The surface of a bone is enveloped in a living membrane called the *periosteum*, which is a fibrous sheet of connective tissue that houses a network of nerves and blood vessels. It is also capable of forming new bone in the event of a break, as long as the circulation is still intact. Inside the periosteum is the hard part of bone, called the *compact bone* because of its dense matrix. Living cells called *osteocytes* secrete the matrix, which is composed of flexible collagen fibers embedded in inflexible calcium compounds such as calcium carbonate. The collagen allows the bone to bend; the calcium keeps it from breaking.

Large bones such as the femur have an active central cavity that contains either yellow or red marrow. In an infant, they contain red bone marrow; in an adult, they contain yellow bone marrow. Yellow marrow is mostly stored fat that can be activated when energy is needed. It can also convert to red bone marrow in stressful times to assist in creating red blood cells, the function of the red bone marrow.

How does the skeleton hang together? If the skeleton were one bone, there would be no need for connective tissue. However, movement would be almost nonexistent. Fortunately, the skeletal structure is made in segments to allow functional movement, which creates the need for bone-to-bone connectors and bone-to-muscle connectors. Where bones intersect, a tough layer of cartilage surrounds the bone tips and protects them from friction. An exception to this rule is where two bones join as a suture, such as the individual bones that create the human skull. In addition, high-pressure areas (such as the hip socket) and high-use areas (such as fingers and wrists) have specialized cells that secrete *synovial fluid* as a lubricant into the joint area to further minimize the wear and tear and friction.

Bands of connective tissue called *ligaments* hold the bones together in proper orientation and conjunction so joints function correctly. Likewise, skeletal muscles are connected to the bone by fibrous connective tissue called *tendons* (more on this in the section "Muscle-Bone Interaction"). *Muscles* may also connect directly into the periosteum of the bone.

Muscles

Muscle cells are specialized cells that are able to contract. Muscle tissue constitutes approximately 50 percent of a normal adult body mass. Muscle tissue is found almost everywhere in the body from operating at the skin level to surrounding blood vessels. The size of an individual's muscles have a DNA linkage, but are mostly determined by an individual's lifestyle. The size and location of each individual muscle determines how well we can perform a task and determines the shape of our body!

Muscle Tissue Types

All muscle tissue is similar in that it is all made from muscle cells. Muscles are the most abundant tissue in virtually all animals. There are three main types of muscles each of which has a slightly different structure and function in the body: cardiac, skeletal, and smooth

Cardiac muscle is the contractile tissue that composes the heart. Cardiac muscles are striated, meaning they appear to have stripes when viewed under a microscope. They are not under the direct control of the organism. Unlike most muscle cells, cardiac cells contract without direct stimulation by the central nervous system. The cells contain only one nucleus, located near the center of the cell. Cardiac muscle cells branch and are interconnected with branching fibers that allow nerve impulses to move from cell to cell during the heartbeat and are tightly packed together with a very narrow synapse so that neighboring cells contract at the appropriate time.

Skeletal muscles are under voluntary control and they are attached to bones by strands of tough fibrous tissue called *tendons*. Skeletal muscle cells are multinucleated, long and slender. The muscle cells, connective tissues, circulatory and nervous system interventions interact to form a functional skeletal muscle.

Skeletal muscles create the force that pulls the bones, which gives movement. They normally work in opposition; for instance, when your biceps muscle contracts, it pulls the forearm and the arm bends at the elbow joint. The biceps is an example of a *flexor* muscle, one that bends a joint and decreases the angle. The triceps muscle is an *extensor* that contracts and acts in opposition to the biceps and straightens the elbow by pulling the forearm straight, moving the forearm back the other way and increasing the angle.

Smooth muscles lack striation and are usually involuntary. They contract more slowly than cardiac or skeletal muscles, but are capable of sustaining the contraction for a longer time period. These spindlelike muscle fibers line the walls of blood vessels, digestive tract, and urinary bladder. They are capable of producing the slow, smooth contractions that move substances such as food through the human digestive system.

Muscles only work when they contract. Muscles are often placed in opposition to each other so that one muscle moves a bone forward while its antagonist is able to move it backward. Sometimes muscles work to provide the unidirectional movement of a substance. The corresponding mechanism of muscle contraction focuses on the striated units of the muscle. In the early 1950s, the British Nobel laureate A. F. Hurley developed the *sliding filament model* after careful observation (using an electron microscope) of striated muscle.

Sarcomere Structure

The process of muscle contraction can be explained with an understanding of the structure of a functional muscle unit or *sarcomere.* A sarcomere is the contractile unit of the muscle and is the area between the striations. Each sarcomere is composed of thin actin filaments and thick myosin filaments that slide past each other. Refer to the illustration *Sarcomere.*

The two lines are proteins that connect adjacent thin filaments and ensure their proper orientation when at rest or during contraction. The area between the two lines is the sarcomere. When a muscle contracts, the sarcomere shortens because of the interaction of the actin and myosin filaments pulling the two lines closer together. Specifically, many small projections in each myosin filament form cross-bridges by connecting with an actin filament. When the muscle is stimulated to contract, the cross-bridges pull the actin and myosin filaments past each other, thus shortening the sarcomere.

Bioterms

Sarcomeres are the functional units of muscular contraction and are composed of thin filaments of actin and thick filaments of myosin. Contraction shortens the sarcomere, which may contract the entire muscle to about one half of its resting length.

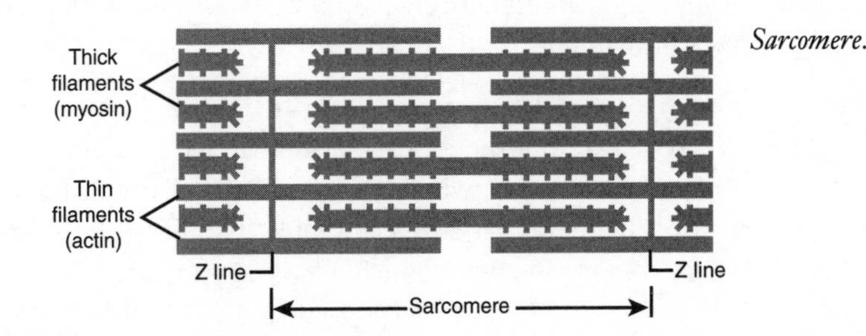

Sarcomere.

Sliding Filament Model and Muscle Contraction

With an understanding of the structure of a sarcomere, the sliding filament model can be explained in four steps. In general, a muscle works by contracting as the filaments slide past each other to shorten the sarcomere. Energy is expended in the process, so working muscles contain hundreds of mitochondria that produce ATP by cellular respiration. Refer to the illustration *Muscle contraction.*

Muscle contraction.

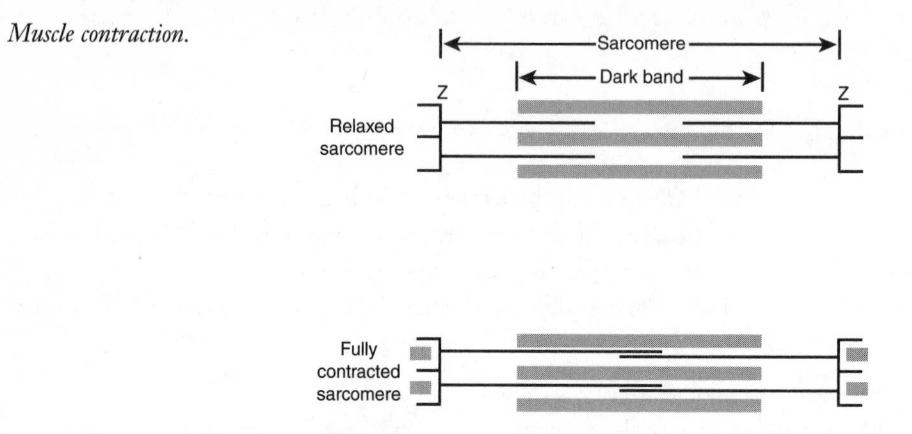

Beginning with the muscle in its resting or noncontracting orientation, the muscle receives a signal to contract. The following four steps refer to only one active site, with the understanding that there are thousands in each sarcomere:

1. ATP binds with a myosin molecule (thick filament) causing it to release from its bond with an actin molecule (thin filament).
2. The ATP releases energy to form ADP, which remains attached to the myosin. The release of energy causes the myosin-ADP complex to reconfigure so the myosin is bent forward in a receiving or "cocked" position for binding with another actin molecule.
3. Calcium ions bind with the actin molecule to create a bonding site for the myosin filament.
4. The myosin and actin fibers unite at the new location and the ADP molecule is released. Without the ADP molecule attached, the myosin reconfigures to its normal resting state, which pulls the actin filament forward. When that happens the thin filaments slide past the thick filaments, the Z lines come closer together, the sarcomere shortens, and the muscle contracts!

The mechanism of muscle movement is set up when the myosin molecules are reconfigured or leaned forward in preparation for movement when they bind with ATP. That forward-leaning posture is reversed when the ADP is released from the myosin-ADP complex.

Because the myosin is no longer reconfigured or leaned forward by the presence of ADP, it returns to its normal state and brings the newly bonded actin molecule forward with it.

Nervous Control of Muscle Contraction

Muscles must work in a controlled manner or they are useless—or worse, harmful. Nervous connections between the central nervous system and muscular system normally provide coordinated muscle movement. For instance, motor neurons, called *effectors*, connect the central nervous system to the skeletal muscles intersecting at sites called *neuromuscular junctions*, while action potentials (see Chapter 15) from the motor neurons regulate the coordinated contraction of these muscles. A typical motor neuron is highly branched and can stimulate multiple muscle cells at the same time. Sarcomeres do not contract on their own.

The intersection of the nervous system with the muscular system regulates muscle contraction. The mechanism of muscle control is described in four steps:

1. Motor neurons release acetylcholine molecules from their axons. The acetylcholine neurotransmitters diffuse across the synapse.
2. The acetylcholine changes the permeability of the receiving muscle cell membrane, producing an impulse that continues the action potential in the muscle.
3. The action potential stimulates the release of calcium ions from the sarcoplasmic reticulum in the cell.
4. The calcium ions then prepare the actin molecule to bind with a new myosin molecule, which causes the muscle filaments to slide past one another.

Bionote

All animals have skeletal muscles organized to work in opposition—even an ant!

The muscle cell remains contracted until the production of acetylcholine is stopped. The neurotransmitter junction produces acetylcholinesterase, an enzyme that degrades acetylcholine. When the acetylcholine is removed, the muscle cells reabsorb the calcium ions and the muscle concentration is finished; the muscle is then in a relaxed state.

The interaction between nerves and muscles is similar to the nerve-nerve interaction. A muscle contracts because a nerve cell stimulates a muscle, which then changes the permeability of the muscle cell membrane causing calcium ions to be released. The calcium ions perform their function and return to the muscle cell to end the muscle contraction.

Muscle-Bone Interaction

Skeletal muscles are set up in opposition, so the action of one muscle complements the action of another. Because muscles can only function by contracting, if a muscle contracts to pull a bone in a certain direction, that muscle cannot reverse itself to push the bones back again. Instead, an antagonistic muscle contracts and pulls the bone back to its original position.

Skeletal muscles are connected to bones by tendons. Tendons and muscles act to make the bones work like levers with the bone joints serving as fulcrums. Normally, multiple muscles are connected to each bone, allowing the bone to move in multiple directions as in rotational movement. The typical example of skeletal muscle action is the bending of the forearm at the elbow. When a person's arm is resting with the palms facing upward, the biceps muscle on the top side of the arm between the elbow and the shoulder raises the forearm. The antagonistic triceps muscle is located beneath the upper arm and connects at the scapula (shoulder) and the two bones that bend the elbow, the humerus and ulna. When the triceps contracts, the forearm moves downward.

The normal resting state for most skeletal muscles is partial contraction. This state of partial contraction gives the muscle a firm or tightened feel, called *muscle tone.* Muscle tone provides the body with posture. This posture is the overall appearance of the body including the position of the head, back, and appendages even when they are not in active use. Muscle tone is increased by regular exercise and may be decreased by nonuse. Exercise increases muscle tone because of the increased amount of proteins in the cells and by the increased efficiency of the energy production in the cells, depending upon the type of exercise, such as anaerobic and aerobic.

The Least You Need to Know

- The skeleton provides protection and serves as the structural support for the body and an attachment site for muscles.
- Bones are made stronger by the addition of calcium carbonate.
- Muscles move bones and provide the range of motility exhibited by animals; smooth muscles provide a slow, enduring contraction, which is often used to move a substance; cardiac muscles are only found in the heart.
- The sarcomere is the functional unit of the muscle cell, and the sliding filament model explains how a muscle cell works.
- Muscle contraction is controlled by the nervous system.

Chapter 17

Reproductive System and Embryology

In This Chapter

- How a human embryo develops
- Hormonal influences involving the reproductive systems
- How oogenesis is similar to spermatogenesis
- The mechanism for childbirth
- The characteristics of trimester development
- How a blastula is different from a gastrula

This is the chapter that fills in the parts regarding sexuality that were omitted from the conversations during the ride home on the school bus in middle school. Each system is examined in relation to function and structure.

In addition, child development is studied as a reproductive process that begins with the development of sperm and egg and ends with a crying newborn and a happy family. Each trimester is examined individually with fetal changes noted for those parents who may eventually be in that situation.

Reproductive System

The human reproductive system is an interesting system to study in detail because of the intricacies of the process and the anomaly that it is the only system that, if nonfunctional, the organism can still function; however, the primary goal of continuing the species would be lost. Although the embryonic testes begin producing small amounts of *androgens*, and the ovaries small amounts of estrogen during the six to eight week period of gestation, neither produce gametes until after *puberty*.

Female Reproductive System

The process by which the female presents an egg for fertilization is a dynamic interplay between form and function. Refer to the illustration *Typical female reproductive system*.

A female is born with all the eggs that she will release for fertilization for her lifetime. The process begins with each egg encased by layers of *follicle cells* located in the ovary. The follicle cells protect and provide nourishment for the egg cell. They also produce *estrogen*, the female sex hormone. Approximately every 28 days, *ovulation* occurs, whereby a follicle releases an egg into the *oviduct*, also called the *fallopian tube*, where cilia move it forward to the uterus. What is left of the follicle develops into the *corpus luteum*, which is a specialized tissue that secretes *progesterone* and continues to secrete estrogen. Progesterone stimulates the buildup and maintenance of the uterine lining during pregnancy. If the egg is not fertilized, the corpus luteum degenerates.

Bioterms

The hypothalamus signals the pituitary gland to release hormones that trigger the growth of gonads. **Puberty** is a period of rapid growth and sexual maturation during which time the reproductive system becomes fully functional. At the completion of puberty, both male and female reproductive units are fully developed and functional.

Bionote

An *ectopic pregnancy* occurs when the embryo attaches to a site other than the uterus, usually the oviduct.

Fertilization normally occurs in the first section of the oviduct, creating a zygote that begins cleavage as it continues to move forward in the oviduct until it reaches the uterus, or womb. Once there, it attaches to the uterine lining for nourishment to promote rapid growth as an embryo and then as a *fetus* after about nine weeks, when body structures begin to appear. The uterus has a thick muscular wall and is designed to accommodate fetal development. Its inner lining, the *endometrium*, is the fetal attachment site because of the immense number of networking blood vessels that sustain rapid growth.

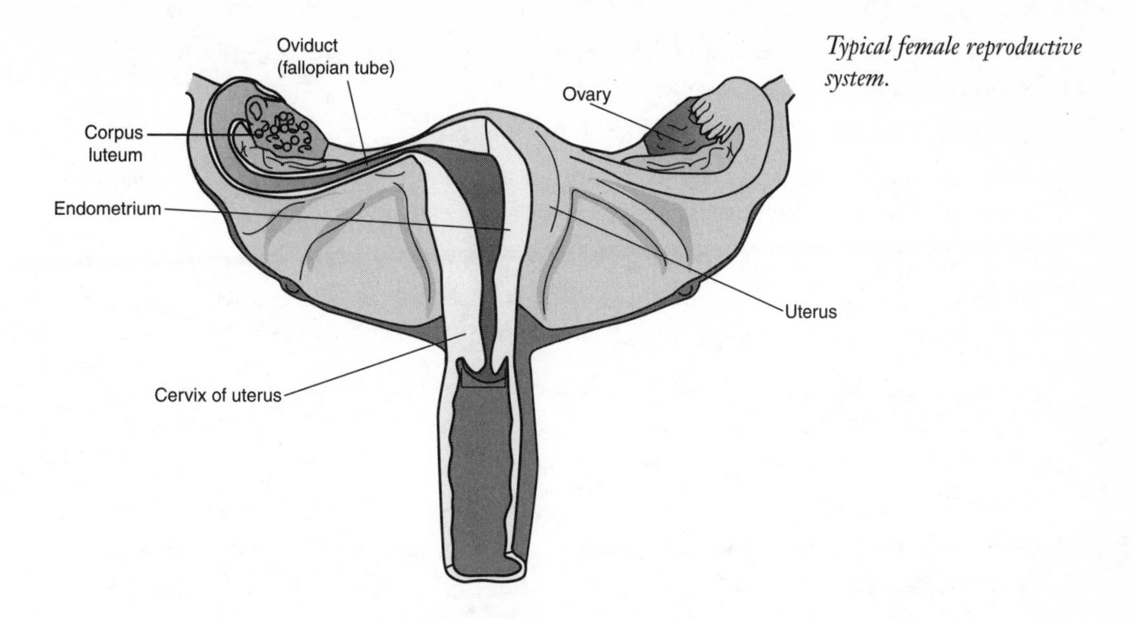

Typical female reproductive system.

The exit and birth route of the embryo from the uterus is through the *cervix*, which must dilate to accommodate the head of the infant as it moves through the vaginal birth canal and into its new world.

Oogenesis and Hormonal Control

The development of a mature ovum can be described in three phases:

1. Prior to birth, *oogenesis*, the development of eggs, begins when the diploid cells inside the follicle become haploid via meiosis.
2. The haploid ovum cells, called *primary oocytes*, remain dormant in the prophase stage of meiosis (meiosis I) after birth until puberty.
3. At puberty, the hypothalamus secretes a release hormone that stimulates the pituitary to release a hormone called the follicle-stimulating hormone (FSH) approximately every 28 days, which activates one follicle to enlarge and allow the ovum to complete meiosis.

The following process describes oogenesis: The division of cytoplasm is unequal, with one *secondary oocyte* receiving virtually all the cytoplasm, whereas the other cell, called the *first polar body*, receives almost none.

The hypothalamus again secretes releasing hormones to activate a second pituitary hormone, luteinizing hormone (LH), to trigger ovulation, and the secondary oocyte enters the oviduct, also called the fallopian tube, from the ovary. If fertilized, the secondary oocyte completes meiosis II and generates the second polar body and the actual egg, which unites with the sperm to form a zygote.

If the egg is not fertilized, a different series of events takes place, called *menstruation.* There are similarities in the first stages.

Ovulation and the development of the corpus luteum progress in the same way: rising hormone levels and an ovum down the fallopian tube increase the growth of supporting tissue in the uterus in preparation for accepting a fertilized egg.

If unfertilized, the egg passes all the way through the uterus and the corpus luteum begins to degenerate. Degeneration of the corpus luteum releases less estrogen and progesterone into the bloodstream.

As soon as the level of estrogen in the blood falls below a threshold level, the enlarged and unneeded uterine lining detaches and, along with the unfertilized egg, exits through the vagina.

Diminished estrogen flow in the bloodstream triggers the hypothalamus to send a releasing hormone to the pituitary gland that stimulates the production of FSH and LH, and the cycle regenerates.

Male Reproductive System

As the female reproductive system is designed for the production of an egg, the male system is structured to produce sperm. Before birth, the male gonads, or *testes,* migrate out of the body into a saclike structure called the *scrotum.* The scrotum is located outside the body to avoid high body temperatures that inhibit sperm production. The temperature of the scrotum is around 2°C cooler than the body. It also houses the *seminiferous tubules* where the actual sperm is manufactured. Refer to the illustration *Typical sperm.*

A mature sperm has three parts:

- The head is an enlarged section that contains the haploid nucleus and a small membrane-enclosed sac, called the *acrosome,* that contains enzymes that dissolve the protective covering around the egg and allow the DNA to enter the egg cell.
- The midpiece is smaller than the head but provides the energy for movement thanks to tightly packed mitochondria.
- Energy created in the midpiece powers the flagellalike tail that swings back and forth to propel the sperm.

Fully mature sperm migrate from the seminiferous tubules to the epididymus where they are stored while they gain mobility and fertilizing ability. Most of the sperm are transported into tubelike *vas deferens* where they migrate and receive *seminal fluid* from three separate glands. The *seminal vesicles* secrete a clear lubricant that nourishes the sperm. The *prostate gland* secretes the opaque, milklike fluid that neutralizes the acidity of any residual urine in the urethra and the natural acidity of the vagina. The *bulbourethral glands* secrete only a minor amount of fluid into the urethra to lubricate it and provide for the easy flow of sperm through it. The sperm in combination with the added fluids is now called *semen*. Refer to the illustration *Typical male reproductive system*.

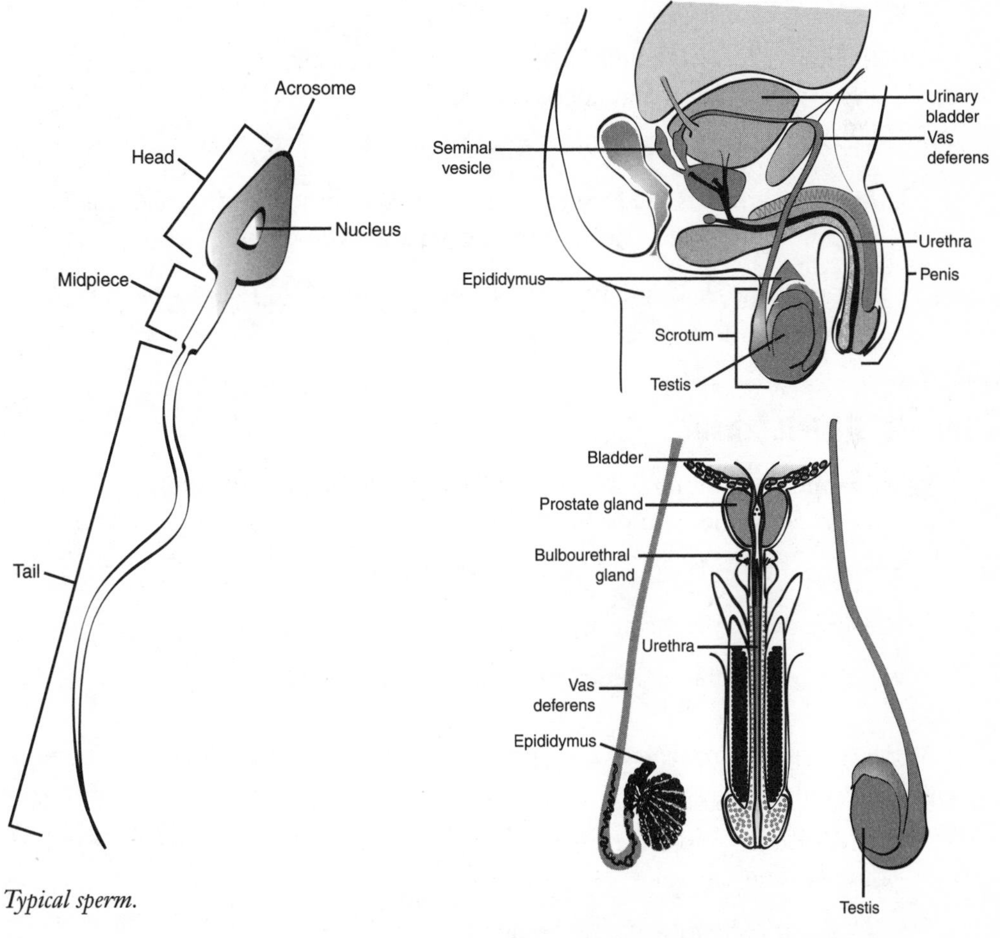

Typical sperm.

Typical male reproductive system.

When the male becomes sexually aroused, the autonomic nervous system (ANS) rhythmically contracts the smooth-muscle lining of the vas deferens, forcing the sperm into the urethra where it joins semen produced in the other testis to be discharged out of the penis. Approximately 250 to 350 million sperm are released during an ejaculation. The overproduction of sperm is designed to make fertilization happen on the first occasion.

Spermatogenesis

The manufacture of sperm normally takes about 70 days and can be described in four steps:

1. Sperm begin as diploid cells that reproduce mitotically to become *primary spermatocytes* that then undergo meiosis.
2. Meiosis I divides the diploid primary into two haploid *secondary spermatocytes.*
3. Meiosis II divides the secondary spermatocytes into four haploid, single chromosomes that then become four separate and unique sperm cells.
4. The sperm cells then pass into the epididymus where they are equipped and stored.

Trimester Development

Within a month of fertilization, the embryo has developed significantly and exhibits unique characteristics of most vertebrates:

- Notochord and coelom both formed from the mesoderm
- Amniotic sac that protects and creates a fluid environment
- Gill pouches that develop into throat and middle ear

At the end of the *first trimester*, numerous changes have occurred that distinguish the fetus as human:

- The placenta and fetus are attached by an *umbilical cord* that contains two arteries and one vein.
- All organs are present in a developing state.
- Arms with hands and fingers; legs with feet and toes are present and can move.

By the end of the first trimester, the zygote looks like a human, including its sex type.

The *second trimester* is a time of rapid growth, unlike the differentiation of the first trimester. The corpus luteum degenerates as the placenta assumes fetal-development responsibilities and begins secreting progesterone. By the end of the second trimester, several structural changes have occurred:

- The fetus has full facial features, including open eyes and the beginning of tooth formation.
- The skeleton begins to form.
- A heartbeat and increased movement.
- A layer of fine hair grows.
- The fetus rotates into the "fetal position," and the mother's midsection is noticeably enlarged.
- Tissues become more specialized and complex.

The *third trimester* continues the rapid growth begun in the second trimester as the fetus completes the final changes prior to birth:

- Circulatory and respiratory organs, especially the lungs, develop rapidly in preparation for open-air breathing.
- The bones harden, except for the head.
- Muscle mass increases, but body movement slows as the fetus fills almost all the available uterine space.

When the fetus is fully developed, usually in about nine months, the process of childbirth begins.

Childbirth

Childbirth involves the interaction of hormones with the female anatomy and can be described in eight steps:

1. Estrogen secreted from the placenta increases the level of hormones in the mother's blood to its highest concentration several weeks before childbirth. High estrogen levels trigger the development of *oxytocin* receptors in the muscle tissue of the uterus.
2. Both fetal cells and the mother's pituitary produce oxytocin, the hormone responsible for the rhythmic contractions of the involuntary smooth muscles in the uterus and the stimulation of the placenta to release *prostaglandins.*

The presence of the prostaglandins causes the uterine muscles to contract even more. The positive-feedback mechanism where the hormone combination stimulates the uterine muscle in turn stimulates the release of more hormones, thus multiplying the effect. Uterine contractions, called *labor*, serve to begin the opening of the cervix in incremental stages to approximately a 4-inch or 10-centimeter diameter that is large enough for the infant's head to pass through.

3. Gradual dilation of the cervix is the longest part of childbirth and may last hours to days. As muscular contractions continue, the baby is moved headfirst toward the cervix that continues to expand.
4. As the baby is pushed toward the cervix, the amniotic sac breaks and the amniotic fluid flows out of the vagina.
5. The baby is then forced through the cervix and out of the vagina, with umbilical cord still attached. If the baby is not instantly breathing, the attendant will usually spank the child on the rear to create a gasp of inspiration that precedes the cry.
6. The blood supply stops flowing in the umbilical cord, making it useless, so it is severed and becomes the only scar that all humans have in the same location, the "belly button." Usually within a half-hour, the remainder of the placenta and the empty amniotic sac are expelled through the vagina.
7. Without the placental hormones, the levels of estrogen and progesterone return to a normal state, permitting the uterus to also return to normal. Decreased progesterone in the maternal blood stimulates the pituitary to release prolactin, which stimulates milk production in the mammary glands.
8. Days later the combination of prolactin and oxytocin causes the secretion of milk by mammary glands. Interestingly, with regular breast-feeding the milk is always available, yet the production ceases when the mother decides to wean the child. In typical interaction between the nervous and endocrine system, physical stimulation of the nipple area creates an impulse from those nerve cells to the hypothalamus, which then stimulates the pituitary to increase the production of prolactin. The converse is also true: The lack of stimulation decreases prolactin production.

Immediately following birth, the newborn's lungs expand for the first time to take in air. The baby usually then cries or is made to cry to help the child clear the air passage and breathe. The umbilical cord is then tied off and cut, completing the severance from the maternal life-supporting system. The umbilical arteries and veins seal off within 30 minutes after cutting. Within minutes the newborn's circulatory, respiratory, and digestive systems are operational, beginning the independence of the individual.

Embryology

The evolutionary history of an organism is most noticeable in the embryonic stages. Humans, like other vertebrates, develop a tail, pharyngeal slits, and limb buds during their embryonic development. In humans, the tail is reabsorbed before birth, but most vertebrates retain their tails, and the pharyngeal slits become gills in fish. Embryonic analysis reveals homologous structures are more apparent. For instance, the embryonic limb buds become chicken wings and human arms. The study of the embryologic development of organisms is called *embryology*.

Embryonic Development

The process of embryonic development occurs in four distinct steps:

1. Once a zygote is conceived, rapid mitotic cell division occurs, called *cleavage*, which doubles the number of cells with each division. Interestingly, the cells become smaller, so the overall mass of the solid sphere of cells does not change! Gene transcription is inhibited while DNA replication, mitosis, and cytokinesis create new cells.
2. Cleavage continues to generate new cells and a blastocoel forms in the center of the embryonic sphere. When cleavage ceases, the blastocoel, which is a fluid-filled area surrounded by cell layers, is reconfigured into a hollow ball of cells, called the *blastula*.
3. During *gastrulation* the blastula begins to unfold and cells migrate to form the three primary germ tissue layers: endoderm, mesoderm, and ectoderm. The endoderm and mesoderm migrate to the center of the cell at a location called a *blastopore;* while ectoderm cells surround the gastrula.
4. Migrating endoderm cells produce a digestive cavity in space previously held by the blastocoel, while the mesoderm occupies the area between the endoderm and ectoderm. Acoelomates and some coelomate animals, such as annelids, mollusks, and arthropods—that develop their mouth near or from the blastopore during gastrulation—are called protostomes. *Deuterostomes*, such as echinoderms and chordates, develop their anus near or from the blastopore, and the mouth forms later.

Bioterms

The germ tissue layer or **germ layer** is a specific arrangement of cells in an embryo from which specific organ systems are derived.

After the gastrula has separated the three *germ layers*, each layer develops organs and organ systems according to its DNA plan. Ectoderm forms the skin; mesoderm, most organs and systems; endoderm, epithelial lining and the remaining organs and organ systems, including the reproductive system.

Cleavage, which creates a multicellular animal; blastulation, which creates a hollow ball of cells; gastrulation, which configures the embryo into three distinct layers; and finally, the subsequent organ development—these are basically the same in all animals. Some scientists believe this is evidence of a single common ancestor.

Human Embryonic Development

Human embryonic development proceeds through similar stages. Cleavage in humans begins in the embryo with 24 hours of fertilization. By the seventh day, cleavage has created a hollow sphere of about 100 new cells that create the child. Cleavage also creates an outer layer, called the *trophoblast*, that integrates with the uterine wall lining and eventually forms part of the *placenta*. The embryonic placenta nourishes the fetus, removes waste, and transmits antibodies from the mother. Gastrulation occurs in about nine days, and the process of differential, specialized growth begins.

Of the many intricate and complex systems of the human body, reproduction is the only one that requires interaction with another human body to successfully complete its function. Under optimal conditions, propagation of the human species is cause for joy and celebration!

The Least You Need to Know

- The endocrine system is involved in every aspect of human reproduction.
- Fertilization normally occurs in the first section of the oviduct.
- Oogenesis creates one egg for fertilization; spermatogenesis creates four sperm.
- At the end of the first trimester, a human embryo has enough observable characteristics to confirm it as a human.
- In human reproduction, the nervous system causes the endocrine system to produce a hormone that is used by the reproductive system.
- The initial mitotic divisions of the human zygote take place in a fallopian tube; after 6 to 8 days the embryo attaches to the wall of the uterus for the remainder of the development time.

Chapter 18

Circulatory System

In This Chapter

- The functions of the circulatory system
- Advantages and disadvantages of a closed and open circulatory system
- The structure of blood vessels affects their function
- The components of blood and their function
- How blood flows through a human heart and circulatory system

The circulatory system is a big chapter. Not because it is long and complicated, but because almost everything has some type of circulatory system. Plus the variation between systems is unique and interesting! It all depends upon what the organism needs to survive in its specific location and what capabilities the organism needs to survive. The circulation of the various fluids changes as animals become more complex. Like the nervous system, the circulatory system reaches every cell for more advanced animals, but not for the simple ones.

Circulatory System

The function of a circulatory system is to provide nourishment, water, hormones, and other metabolic requirements to each cell in a body. Likewise, the circulatory system removes metabolic wastes from cells and transports them to the appropriate area or organ for processing.

A circulatory system is a requirement for advanced animals because the process of diffusion is too slow for their active lifestyles. The rate of diffusion is generally the controlling factor in how large an animal without a circulatory system can grow. The inefficiency of diffusion prohibits animals from diffusing substances across areas greater than three to four cells with the speed necessary to support an active metabolism. In all cases, the circulatory system is responsible for feeding and cleaning every cell in the body.

Invertebrate Circulatory Systems

Unicellular and simple animals have their body cells located near or interfacing with the external environment. Because each cell is capable of interacting with the environment whenever needed, there is no need for a circulatory system. In fact, both Porifera and cnidarians lack a circulatory system. The cnidarians utilize their gastrovascular system to help transport substances much like a circulatory system.

Mollusca and Arthropoda Circulatory Systems

Both molluscs and arthropods have an open circulatory system. This advancement allows these animals to grow larger with more cell layers in depth because they do not have to interface directly with the environment.

An open circulatory system is less sophisticated than a closed system. In an open system, the heart pumps blood through blood vessels that terminate in a body cavity. The blood floods all of the cells and tissues located in the cavity, thereby processing their needs. The blood pool in the body cavity feels little if any pressure from the heart, so the rate of flow is very slow. This slow rate is beneficial for the exchange of gases, nutrient acquisition, and water-balance maintenance. Muscular contractions send blood into returning vessels that carry the blood to an area of gaseous exchange like a gill. The oxygenated blood is then moved to the pump or heart which then sends it back to the body. In an open system, the blood is in direct contact with the cells.

Molluscs have a three-chambered heart that pumps oxygenated blood into its finger-like cavities, called *sinuses*, where blood is in direct contact with all interior cells and tissues. The sinuses also channel the movement of blood through the gill area for gaseous exchange. The blood is then transferred back to the heart for redistribution to the body. An open circulatory system should not infer the meaning of random, disorganized flow. Although it is not as directed or efficient as a closed system, it is adequate for sessile or low-metabolizing animals. It also does not require the presence of an *interstitial fluid* that is needed in a closed system.

Arthropods, such as the common crayfish, have a heart located on the top or dorsal side of their body. The heart pumps a bloodlike substance called *hemolymph* into blood vessels that transport it to the *hemocoel*, which is a centralized body cavity, and associated sinuses. While in the hemocoel, the hemolymph floods each tissue and is then pushed into the gill area for gaseous exchange. The hemolymph receives oxygen and eliminates its carbon dioxide load before returning to the dorsal heart. The "flow" of the blood is somewhat circular in nature. The heart pumps it into the hemocoel, which is normally in the central area on the ventral or bottom side. Muscular contractions and body movements move the blood in a return cycle to the dorsal side for repumping. Interestingly, molluscs, arthropods, and other coelomate animals have a mesoderm layer around their digestive tubes that prohibits diffusion! Coelomates had to develop a circulatory system.

Bionote

Squid and octopuses are an exception to the open circulatory system of molluscs and arthropods. Their lifestyle and subsequent high metabolic rate necessitate a closed circulatory system.

Annelida Circulatory System

Annelida is the first animal phylum to exhibit a closed circulatory system. Near the head or anterior end of most annelids, such as on earthworms, muscle tissue surrounds some of the blood vessels and contracts rhythmically to push the blood forward in the vessels. These specialized blood vessels are called aortic arches, and their muscular pumping serves the function of the heart and moves the blood along two blood vessels that run the length of the worm. The blood moves in a circular pattern: On the dorsal side, blood is returning to the pumping area; on the ventral side, blood is leaving the pumping area to return to the body. Circular blood vessels connect the two major blood vessels so that the blood can create a small loop within each body segment as well as a large loop around the entire organism.

A closed circulatory system confers many advantages. The greatest advantage is the efficiency that accompanies the unidirectional, organized flow. It also provides adequate sustenance to all cells to support a high-metabolic lifestyle. In a closed system, each cell is guaranteed to get the service it needs, plus the system can regulate itself by temporarily opening and closing various areas to shunt blood flow to areas of greater need. Interestingly, a closed system does not bring the blood into direct contact with the cells. Substances that move in and out of the blood vessels must move through the interstitial fluid before entering. The interstitial fluid surrounds each cell and provides an interface for the movement of molecules.

In a closed circulatory system, the flow of blood is totally contained within three types of blood vessels: arteries, veins, and capillaries.

Arteries usually carry oxygenated blood away from the heart, with one notable exception. In a double-loop circulation pattern, reduced-oxygen blood from the body is returned to the heart, which pumps it through *pulmonary arteries* to the gaseous exchange organs such as human lungs. The pulmonary artery carries deoxygenated blood to the lungs, and the *pulmonary vein* carries oxygenated blood from the gaseous exchange location back to the heart for pumping back in to the rest of the body. Refer to the illustration *Artery.*

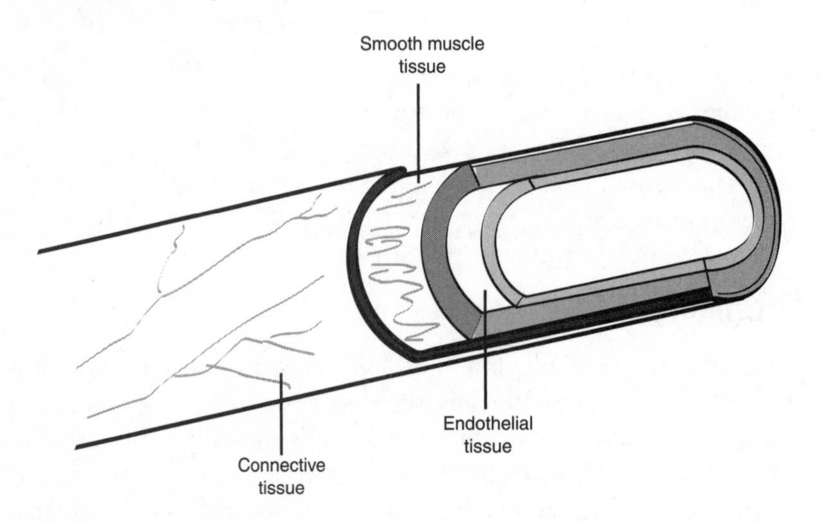

Artery.

Capillaries perform a vital function that is supported by their unique structure. They are the smallest blood vessels in diameter. They also have the thinnest surrounding wall. Refer to the illustration *Capillary.* The capillaries are responsible for allowing the movement of materials into and out of the circulatory system, so their thin, porous membrane and increased surface area are well suited for their function. Refer to the illustration *Capillary. Veins* return deoxygenated blood to the heart except in the noted pulmonary system. Refer to the illustration *Vein.*

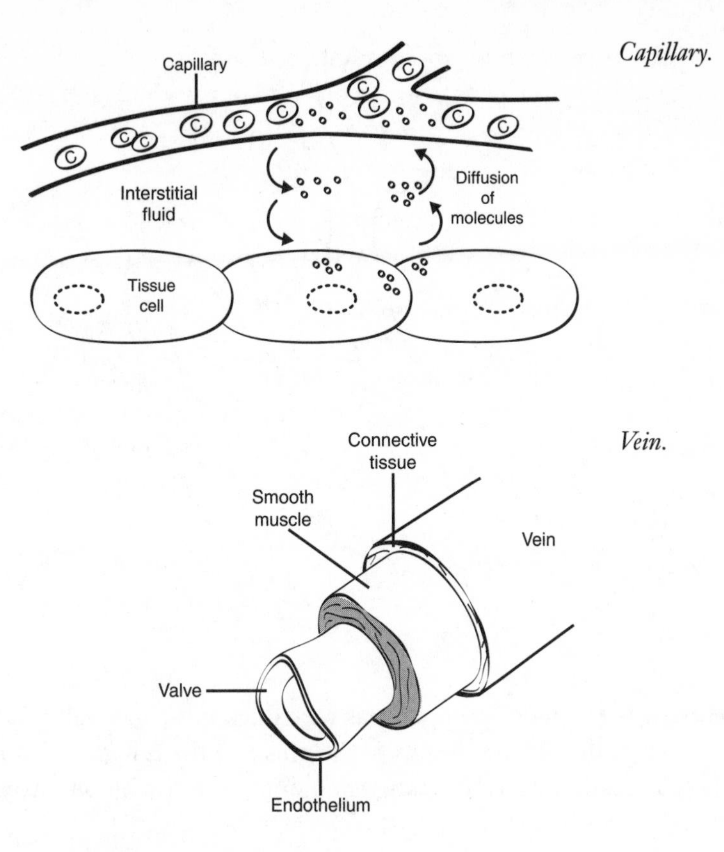

Capillary.

Vein.

Fish Circulatory System

The circulatory system in fish exhibits several advancements. The major advancement is the development of a more efficient heart. In simple animals, muscle contractions around a blood vessel created a "tubular heart" that was sufficient for open circulatory systems. With the advent of more active lifestyles and high metabolic rates needed for support, a more powerful and efficient heart is required. Fish have responded with a modified four-chambered heart that contains two enlarged blood vessels as two of the chambers. The mechanism of action can be described in four steps. Refer to the illustration *Fish heart.*

Oxygen-poor blood returning from the body is collected in the *sinus venosus.* This chamber collects and stores the blood prior to entering the heart in the correct orientation. It also stabilizes the blood pressure of blood entering the heart.

Blood passes directly into the *atrium* from the condensing sinus venosus tubes. The atrium is the holding area that delivers blood to the ventricle. The atrium is expansive and has a thin layer of muscles that push the blood into the ventricle.

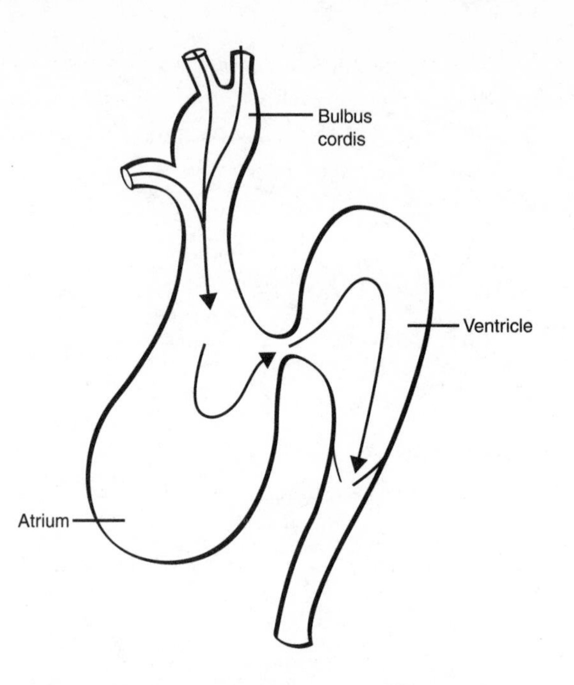

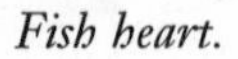
Fish heart.

The ventricle is the actual pump. It has very thick muscular walls that forcibly contract and propel the blood out toward the gills and the remainder of the body. This muscular contraction must be strong enough to push the blood through the entire system!

The *conus arteriosis* is a second pump that receives blood from a ventricle and pushes it forward to the gills. It has muscular walls that are elastic to handle the glut of blood coming from the ventricle. The conus arteriosis expands to accept the entire load and then pushes the load adding even more force. The initial collection of blood from the ventricle establishes a rhythmic, smooth pulse of blood as opposed to erratic pulsations that might create boom-or-bust traffic jams farther down the line.

The fish heart is a major advancement because it pumps blood to a gaseous exchange point and then sends it in a single loop around the body.

Amphibian Circulatory System

Amphibians have a circulatory system that extends the capabilities of the fish system most notably. Amphibians exhibit a double-loop circulatory pattern. This double-loop pattern provides a better supply of oxygenated blood to support increased muscular activity and is continued in more advanced animals such as man. The double loop is an advancement because the blood is pumped to a gaseous exchange point, such as the lungs, and is then returned to the heart for pumping to the remainder of the body. In this way, the blood pressure remains high and the organism can be more active and grow to larger sizes. With the single-loop system, blood pressure was lost as the blood passed through the narrow gaseous exchange locations and never had a chance to regain its pressure. Refer to the illustration *Single- and double-loop circulation.*

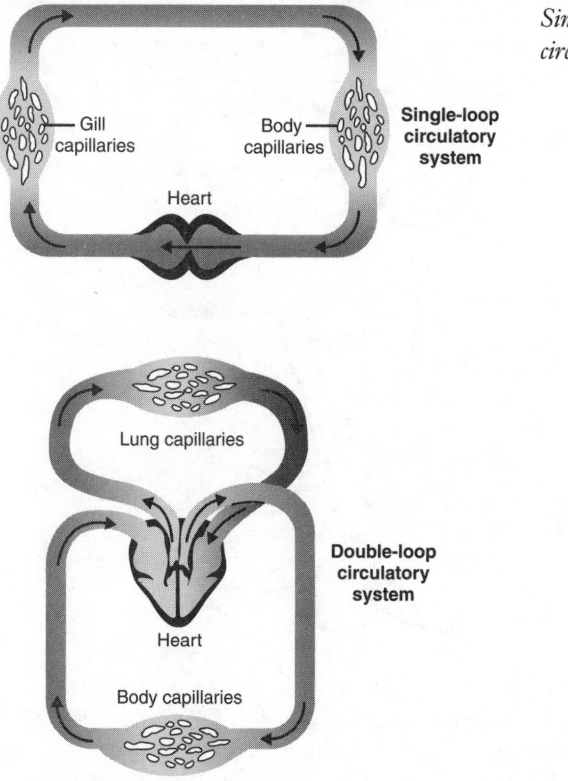

Single- and double-loop circulation.

The structure of the frog heart is another advancement in overall circulation capabilities. Refer to the illustration *Frog heart.*

The frog heart is a three-chambered heart with a dividing wall, or septum, between the right and the left atria. The septum prevents the intermixing of blood between the right and the left atria of the heart. The flow of blood through the frog heart is a typical example of a double-loop circulation pattern. It can be described in three steps:

1. Deoxygenated blood returning from the body enters the right atrium. Oxygenated blood returning from the lungs via the pulmonary vein enters the left atrium.

2. The left atrium empties oxygen-rich blood into the single ventricle. The right atrium empties oxygen-deficient blood into the same common ventricle. The ventricle therefore has a mixture of oxygen-rich and oxygen-poor blood.
3. The muscular ventricle contracts and sends a mixture of blood to the lungs and to the rest of the body.

Frog heart.

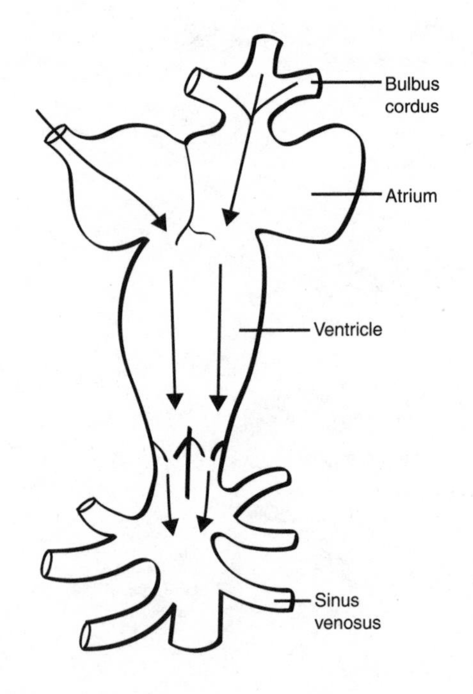

The heart is shaped so that a slight physical separation often occurs between oxygenated and deoxygenated blood in the ventricles; however, a mixture is pumped out. Fortunately, amphibians are also able to compensate for the blood mixture by absorbing oxygen directly through their skin.

Reptile Circulatory Systems

Reptiles are usually more active than amphibians and are unable to absorb oxygen through their skin, so they require a more sophisticated circulatory system. The major advancement is the continued separation of the heart. The septum is extended farther within the ventricles to partially separate the oxygen-rich from the oxygen-poor blood. Although the septum is not complete, it does minimize the intermixing of the blood. Refer to the illustration *Reptile heart.*

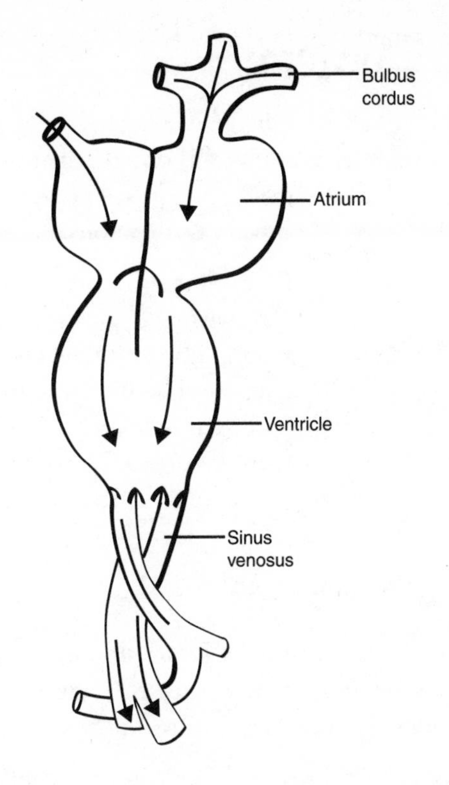

Reptile heart.

Essentially, the reptilian heart works very much like an amphibian heart. The advantage of the reptilian system is the improved quality of oxygenated blood departing for use by the body.

Interestingly, the reptiles take an unusual advantage of the incomplete septum. A reptile can conserve energy by deliberately bypassing the pulmonary system. If the animal is sleeping or inactive, the metabolic level is so low that normal gaseous exchange is not required, so they shunt the blood flow increasingly away from the lungs until needed. Bypassing the lungs also serves another purpose. It can warm the body in an emergency for quick action. Warm blood from the outer sunbaked layers can be shunted into vessels that carry the warm blood to the internal organs! This type of circulatory redirection is not found in more advanced animals, including man.

Bionote

Alligators and crocodiles are an exception to the rule that reptiles only have a partial septum. They have a complete septum that creates a four-chambered heart.

Avian Circulatory System

Birds are the first endothermic animals that experience a continual drain on their circulatory system. An endothermic animal is one that has a high metabolic rate and generates large amounts of internal heat. They also expend tremendous amounts of energy in flight.

Bionote

A typical hummingbird's heart beats at approximately 600 bpm to support its unusual and demanding flight pattern. A typical human heart beats at about one tenth of that rate.

Except for the crocodilians, birds are the first animals with a four-chambered heart. The complete septum divides the ventricles so that blood types do not mix. The improved quality of oxygenated blood is needed to support their energetic lifestyle. The four-chambered heart of the common bird works very much like the action of the human heart.

Human Circulatory System

The circulatory system in humans is based on structures present in simpler animals. In general, the overall function of the circulatory system remains the same. In humans as in other animals, there are four main functions:

- Exchange of oxygen and carbon dioxide at the cellular level.
- Digested nutrients are distributed to all cells.
- Distribution of heat for endothermic animals.
- Transport of hormones from their manufacturing location to their point of action location.

Circulatory Pathway in the Human Heart: Systole and Diastole

The human circulatory system is totally dependent on the continued function of the human heart. Fortunately, much is known about how the heart works and the flow of blood through the heart. The human heartbeat is called a cardiac cycle and is essentially two separate functions known as the diastole and systole.

Diastole typically lasts about 0.4 seconds in normal human hearts. Essentially, two events occur during diastole:

- Deoxygenated blood collected at the *vena cavas* enters the right atrium. Oxygenated blood returning from the lungs via the pulmonary vein enters the left atrium.

- The atrioventricular valves (AV) open allowing blood to flow from both atria into the ventricles.

During diastole, all four chambers of the heart are relaxed and all chambers fill with blood. Both atria drain most of their blood contents into both ventricles before refilling.

Systole also lasts about 0.4 seconds and contains two separate events:

- Slight contraction of the atria forces the remaining blood into the ventricle, 0.1 seconds.
- Both ventricles contract and push blood through the semilunar valves and into the arteries. The left ventricle sends blood via the *aorta* into the body. The force of the ventricle contraction closes the AV valves preventing the backflow of blood into the atria. The contraction of the ventricles lasts about 0.3 seconds.

Medical doctors are trained to listen carefully for the familiar sound that the heart makes. The "lub-dub" is the sound of a normal heart. The "lub" sound is caused when the ventricles contract and slam the AV valves shut. The "dub" is the pressure of ventricular-propelled blood hitting the semilunar valves. Whenever a doctor hears a leaking or swishing sound, it means a possible defect in one of the valves. A heart murmur occurs when a stream of blood is able to squirt back through one of the valves. Although some people are born with heart murmurs, they are often not life-threatening. Rheumatic fever is a disease that affects the heart valves and may create a more serious situation. Damaged valves may have to be replaced surgically with a donor valve or an artificial one.

The heart has an electrical control center that regulates the cardiac cycle so that the two halves work together to provide maximum efficiency. The electrical control of the human heartbeat can be described in three steps. Refer to the illustration *Human heartbeat regulation.*

The sinoatrial node (SA) or pacemaker creates an electrical impulse that spreads through both atria causing them to contract in unison. The signal is received by the atrioventricular node (AV) located between the right atrium and ventricle. The AV node is so dense that it delays the contraction signal by 0.1 seconds. The signal leaves the AV nodes by specialized tissues and triggers the ventricular contractions

Bionote

The 0.1-second delay at the AV node allows the ventricle to contract after the atrium so that they work in a coordinated manner and do not contract at the same time.

at the bottom end of the ventricle first, then moves upward. This allows the ventricle to empty most of its contents.

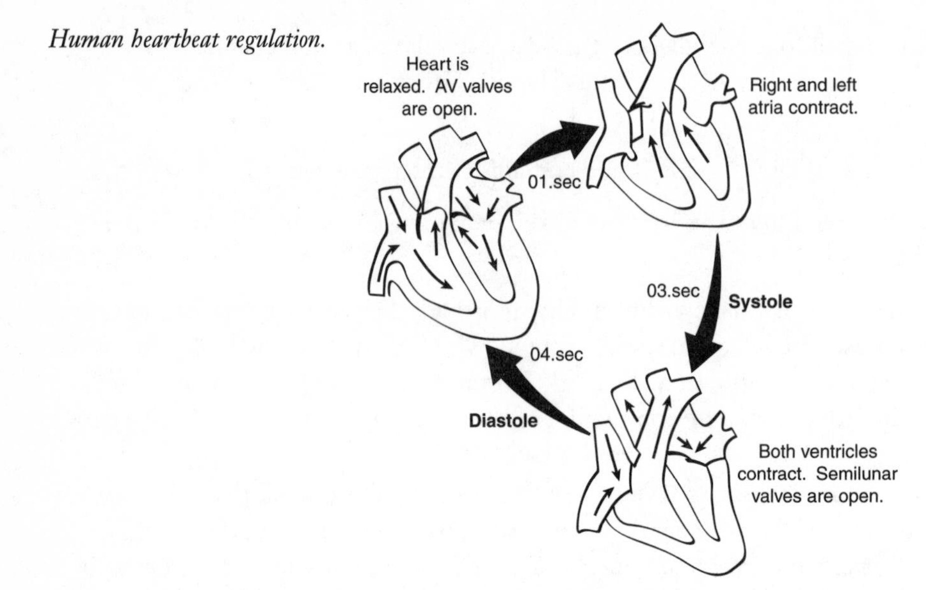
Heart is relaxed. AV valves are open.
Right and left atria contract.
01.sec
03.sec
Systole
04.sec
Diastole
Both ventricles contract. Semilunar valves are open.

Human heartbeat regulation.

The amount of blood that leaves the left ventricle every minute is the *cardiac output.* It can be calculated if the number of heartbeats per minute is known. First, assume that the average human heart expels about 75 milliliters (ml) of blood with each contraction. Most human hearts at rest beat 60 to 80 times per minute (bpm). If we assume 80 bpm × 2.55 fluid ounces (75 ml or .075 liters), the cardiac output is approximately 6.36 quarts or 6 liters per minute.

Human Body Blood Flow Pattern

The flow of blood through the human body takes a series of interesting turns. After the blood leaves the heart from the left ventricle, it circulates through the body before collecting in the vena cavas. The blood flow can be explained in four steps. Refer to the illustration *Human blood flow.*

Oxygenated blood leaves the left ventricle in a glut that stretches the aorta. The aorta is large and flexible and branches immediately to deliver blood to the heart, head, and upper body. The aorta bends behind the heart and branches to deliver blood to the middle organs and lower body.

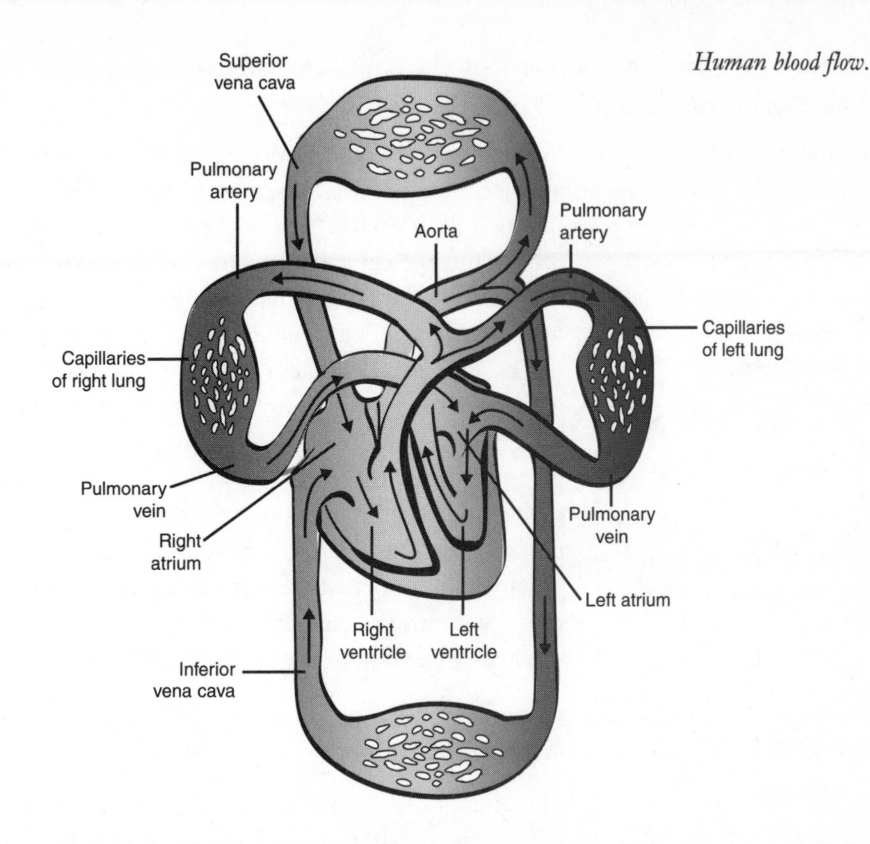

Human blood flow.

The arteries branch into smaller and smaller vessels until they are so small that only one red blood cell can pass through them at a time. Vessels of this size are called capillaries, and they are the functional units of the blood vessels. Refer to the section "Annelida Circulatory System."

On the return trip, the systolic pressure is often depleted because of the massive capillary network. So the action of neighboring muscles and body movement help force the blood into veins that begin collecting the blood into successfully larger veins until the blood in the body is collected in the *inferior vena cava* and blood in the upper half is collected in the *superior vena cava.* Both then empty into the right atrium for holding prior to entering the right ventricle. The pressure of the blood varies as it travels through the body.

Blood Pressure in Humans

Blood pressure in humans is the force that blood applies on the walls of the vessels. It is created when the left ventricle contracts and pushes a glut of blood into the body. This glut of blood increases the diameter of the blood vessel as it passes through, which is easily measured as a *pulse*. Systolic ventricular contractions create the pulse and are easily measured.

Blood pressure drops for several reasons. First, the physics of pushing a liquid through a tube guarantees friction problems caused by the viscosity of the liquid and liquid-wall interface. Second, and more biologically important, the capillaries form a massive network of extremely small vessels that greatly increase the liquid-wall interaction. This reduces the blood speed to a minimum. Fortunately, that is good for the system because it allows time for diffusion and osmosis to occur. It is here that the function of circulatory interface takes place. The blood pressure is so low as the blood leaves the capillary bed that it does not have enough systolic pressure to return it to the heart. As in simpler animals, gravity, body movement, and muscle action help push the blood into the collecting veins for the return trip to the heart. Venal pressure is often very low.

Bionote

A sphygmomanometer is the armband-like device that is used to measure blood pressure. Your blood pressure is recorded in two numbers. In humans, a blood pressure of 120/80 is considered normal. Systolic ventricular contractions are the first number and the diastolic measure is the second number.

Biohazard

A heart attack occurs when the blood supply that nourishes the heart is blocked. When that happens, cells downstream from the blockage do not receive sufficient service, so they die. If enough cells die, the heart becomes nonfunctional.

Human Blood

Depending on the size, the average human holds about 5.3 quarts or 5 liters of blood. The composition of human blood parallels its function. Almost 55 percent of blood is water with dissolved minerals, proteins, and cellular nutrients or waste. This section of blood is called *plasma*. The remaining 45 percent consists of formed elements within the blood that are mostly red blood cells, with a lesser number of white blood cells and platelets.

Red blood cells function to carry oxygen. They are quite small when compared to the large white blood cells, but each has a biconcave disc shape to increase their surface area and withstand changing osmotic conditions. Each cell also contains an estimated 250 million hemoglobin molecules that transport oxygen. Red blood cells usually last about four months before they are recycled by the cell. The element iron is the central atom in the hemoglobin molecule. Iron-deficiency anemia is a lack of iron in the body to make red blood cells. Anemia can also be caused by excessive blood loss, bone marrow cancer, and vitamin deficiencies. Bone marrow cancer causes anemia because red blood cells are produced in the marrow of large bones.

White blood cells or *leukocytes* comprise a group of five different cell types that all work to fight infections. Basophils are probably the most common white blood cell. They release chemicals, usually histamines, which dilate blood vessels so that other white blood cells can leak into the interstitial fluid to combat the infection. Neutrophils, which are the most abundant, and monocytes are usually the white blood cells that pass through the vessel wall and move into the interstitial fluid surrounding the infected body tissue. Neutrophils and monocytes work as phagocytes to engulf and destroy invading microbes. Eosinophils are also phagocytes but they specialize in fighting parasitic worm infections. Lymphocytes are covered in greater detail in Chapter 15. They are capable of producing antibodies and are specialized to attack specific invaders. Together, the white blood cells act to minimize disease and infection within the body. In fact, most of the white blood cells are located outside the blood vessels.

Platelets are the components of blood that are involved in the clotting of blood to prevent excessive bleeding from a wound.

Red blood cells, platelets, and white blood cells are all produced from *stem cells* in the bone marrow. Much research has been devoted to stem cells as a treatment for certain types of cancer. Injection of healthy stem cells into a body can stimulate the regeneration of a totally new blood complement. Research has also examined the possibility of similar treatments for sickle-cell disease and HIV.

The Least You Need to Know

- Circulatory systems in advanced animals provide nourishment, remove wastes, regulate water balance, fight infections, and transport hormones.
- The simplest invertebrates do not have a circulatory system because they can interface directly with the environment.
- Reptiles are able to reroute their circulatory system for metabolic benefit.
- A closed system is more efficient than an open system, but an open system brings blood directly to the cells.
- The evolution of the heart began as tubular pumps in annelids, and advanced through a three-chambered heart in amphibians and a four-chambered heart in birds and mammals.
- Blood is a mixture of substances that protect, build, and maintain the body.

Chapter 19

In This Chapter

- Learning benefits an organism
- Learning and the genetic diversity of a population
- How natural selection affects behavior
- Why animals bluff and play-fight
- Why some animals seem to be born with some knowledge
- Why dogs smell the private areas of other dogs

Have you ever watched the video clips of animals fighting in the wild? Often they are fighting another member of their own species or they are attempting to subdue it for food. Why would an animal want to attack its own kind? Why have humans waged numerous world wars, conflicts, riots, and barroom fistfights? This chapter isn't intended to explain why we don't have world peace, but you will have a better understanding of biological behaviors.

Behavior is genetic, learned, and revised based on group rules. Most social behavior in all animals is designed to prevent injury and make life easier. Even communication is a form of behavior that is designed to assist the living to prevent them from becoming the dying. Migration is a type of social behavior that keeps everyone moving to greener pastures, while the territorial nature of certain animals keeps others off their pastures.

The environment forces a set of behaviors that are generally a do-or-die set of rules. Animals that abide by the rules live and prosper, the others, don't. Often the environmental behaviors are specific to a particular species, such as looking for food. Sometimes different species form a relationship that helps both parties, or helps only one party, or helps one and harms the other. These are all types of behavior that are common in all animals, even some of the ones you are already familiar with.

Behavior

Behavior is an organism's response to a stimulus. How an organism reacts to a situation defines the behavior or behavior pattern for that organism. Behavior patterns are affected by genetics and environmental pressures. Species-specific behavioral patterns are common. Different species living in the same area often have similar behavioral patterns because of similar environmental pressures. An animal's behavior is a reflection of the evolutionary fitness of that species. Natural selection favors behavioral patterns that increase the likelihood of an individual surviving to reproduce. Knowledge is passed from generation to generation within a species in a variety of ways, some genetic. Studying an animal's behavior patterns provides insights into its evolution and ecological interrelationships.

Innate and Instinctive Behavior

Konrad Lorenz and Niko Tinbergen, both Nobel laureates, were among the first behavioral biologists, and their early experiments helped define *innate behavior.* Innate behavior is often called *instinctive behavior* because both are considered an inborn characteristic in that every member of that species acts exactly the same way in a given situation. In the case of a marsupial, the kangaroo, the newborn instinctively searches for and finds the mother's protective pouch without any parental help or prior teaching. Innate behaviors appear to be part of the genetic blueprints for a species. As such, it is controlled by natural selection, which favors those animals that have enough innate knowledge not to spend time and energy learning the fundamentals of their species. Zebras can stand within minutes of birth; they can walk and run within hours. Slow zebra offspring are easy prey, so they do not reproduce.

Bionote

Labrador retriever dogs will return a thrown stick with no prior training.

Lorenz and Tinbergen's work identified a curious behavior that is similar to operating on "automatic pilot." They explained how innate knowledge is expressed in defined absolute behavior expectations that are called *fixed action patterns* (FAP). An FAP is a sequence of events preprogrammed into the individual that once started in response to a stimulus must be completed even if the stimulus is removed. Lorenz's classic example references his work with the graylag goose. To demonstrate an FAP, Lorenz removed an egg from a brooding nest and placed it approximately 3 feet or 1 meter away such that the mother could see it. Instinctively the mother goose arose, extended her neck, and nudged the egg with the side of her head until the egg was back in the nest. This pattern was repeated in different locations with different mother graylags with the same result. To further demonstrate the FAP, when Lorenz swiped the egg while the mother was nudging it back to her nest, she continued the motion even though the egg was no longer there!

Bionote

A human fixed action pattern occurs when a newborn is touched on either cheek. The baby instinctively turns his head and prepares to breast-feed.

It can be proved that DNA codes for certain types of behavior, which is the essence of innate learning. The landmark example of genetics affecting behavior is the study involving the hybridization of two separate species of lovebirds. In one species, the parents create large strips of nesting material, which they transport in their beaks to the nesting site. Another species cuts smaller pieces and stuffs them between back feathers for transportation. A mating between these two species created a hybrid that demonstrated innate knowledge received from both species. The hybrid lovebirds cut long pieces of nesting material and tried to stuff them between back feathers—unsuccessfully!

Learned Behavior

Learning is the act of behavior change as the result of knowledge gained from experience. Five types of learned behavior are exhibited in the animal kingdom: associative, habituation, imitation, imprinting, and reasoning. Most are well known in the education and psychology fields because both fields merge with animal behavioral studies.

Associative

Associative learning is the type that we often learn in school where a specific response receives either a reward or a punishment. Scholastic achievement is rewarded with praise and good grades. Conversely, a child learns to associate the word *no* when playing with matches.

One special type of associative learning is called *classical conditioning*. This type of learning occurs when an abstract stimulus creates an association without the benefit of reward or punishment.

A famous Russian biologist, Ivan Pavlov, created the landmark classical conditioning response experiment when he rang a bell immediately before feeding his dogs. The dogs eventually associated the bell with food. To prove the classical conditioning, Pavlov demonstrated that his dogs would salivate in preparation for eating with only the stimulus of the ringing bell.

Operant conditioning is another type of associative learning that is more common in nature. It is also called *trial-and-error learning* because an animal learns to behave in a certain way based on a previous good or bad experience. As an example, a child learns to avoid a strong heat source after he gets burned. Also predators avoid monarch butterflies because of their bad taste. Usually pet dogs learn to avoid skunks often after only one eventful encounter.

Bionote

Commercial advertising is sometimes based on this principle of associative learning. If a successful manufacturer displays a prominent trademark, then often only a picture of the trademark is needed to remind the consumer of the product. It is thought that once the consumer is reminded of the product, they are more likely to purchase it when shopping.

The classic operant conditioning experiment was conducted by B. F. Skinner, an American psychologist and inventor of the "Skinner box." Skinner placed laboratory rats in the Skinner box, which had a lever that released a pellet of food when pushed. In the beginning nothing happened. Finally one of the rats would hit the bar by accident and receive food. Over time, the rats learned to push the lever to release food whenever they were hungry!

Habituation

Habituation is the easiest type of learning for animals because it is a decreasing response to a repeated stimulus that has no effect and conveys no new information. Actually, the animal learns to ignore the stimulus. Gardeners know that a scarecrow only works for a while and then the invaders get accustomed to it and learn to ignore the stimulus. However, moving the scarecrow to a new location or dressing it differently restarts the habituation clock. Habituation is also a survival advantage because the animals are not bothered by insignificant events, allowing them to concentrate on more important events such as food, mating, and predatory stimulations.

Habituation is hard to overcome and almost always requires a directed conscious effort to override. Failure to override a habituation can be costly. Predatory cats in the African savannah use habituation to spring their ambush. Gazelles, zebras, and other grazing animals must always fight habituation when looking in the monotony of savanna grasses for the cat ready to spring an ambush! The cats know that even sparse cover is sufficient if the prey is seeing but not observing. Police are trained to notice minute details, which is another counter-habituation tactic. Wily animals know when to see and when to observe.

Imitation

When juvenile cheetahs observe their mother in her hunting mode using carefully camouflaged and stealthy tracking techniques and crouched silent stalking in preparation for a successful prey capture, they learn to do it the same way. *Imitation* is copying the actions of a model that has demonstrated success or generated interest. Parents of families with siblings know that the younger children often pick up mannerisms and other behaviors by watching their older siblings. Imitation can simplify learning because the learner gets to see the behavior before implementing the action during a similar event.

Imitation is also a survival strategy because the learner is imitating an organism that is still alive! Even if the organism only learns to repeat what it has seen and not expand on it, it is learning techniques that have allowed the teacher to survive. Bad tactics tend to die with the owner. A drawback to imitation is the limitations imposed by a changing environment. In other words, the strategy for running down a gazelle might not work against a herd of migrating water buffalo.

Bionote

Humans provide an extended childhood for their offspring to devote extra time for their rearing, including time for modeling acceptable, normal, and culturally appropriate, human behavior.

Imprinting

Imprinting is unique because it involves a combination of innate learning and experience. Lorenz again produced the classic experiment. He divided an equal number of fertilized graylag goose eggs into two piles. The control group he left with a brooding mother; he incubated the experimental group in a separate location. Newborns instinctively look for their mother, so when the control group hatched, they looked and found their mother, who they identified with and followed during normal excursions. When the experimental group hatched, they looked and found Dr. Lorenz, who they identified as their mother and followed everywhere!

Other tricks have been played on newborns, such as substituting a wooden facsimile in place of the real mother and noting that over time offspring prefer the facsimile over their real mother! It is not by accident that in most human birthing scenarios, the baby is handed to the real mother as soon as possible. In certain species, such as horses, their unique smell is also a guide for imprinting. Both mother and foal know each other by smell as well as by sight. Imprinting is the first innate behavior to be expressed, and it forms a tight personal bond so strong that it is one of the most difficult to unlearn. This is a survival strategy to keep offspring in close contact with their parents for protection and sustenance.

Reasoning

Reasoning is the capacity to behave correctly in a new situation based on the synthesis of previous related events. Reasoning involves the memory of incidents and the ability to extract the meaning and transfer it to a new situation. This phenomenon is not attributed to any animal groups below primates. Any pet owner who has ever walked an animal on a leash or tethered one to a pole knows how they tangle their leash, your legs, their legs, and any bush that may be near with no apparent understanding of how to prevent this annoying behavior!

It is sometimes called inventive or innovative learning because the individual must create an answer to a problem that it has not encountered before. Aside from humans, chimps are the most widely studied. Most teachers know the research that demonstrates that a chimp will eventually use a stick, or other experimental device, to capture a banana from an out-of-reach location. Similar behavior is often rewarded in corporate America.

Animals that have the ability to learn and remember have a reproductive advantage because they are not as likely to make a fatal mistake. Whether the behavior is instinctive or learned is irrelevant as long as the animal responds appropriately. The ability to reason is a survival factor that helps an organism in a changing environment. The ability to think through a situation is as valuable in the Amazon rainforest as it is in corporate America.

Environment Influenced Behavior

The previous section dealt with genetically controlled innate or instinctive behavior, the interaction of genetics and environment imprinting, and the organism's ability to learn through other methods. This section explores those behaviors that are regulated directly by the environment or by regulatory factors in the environment. Both the *biotic* and *abiotic* factors in the environment influence behavior and the transfer of behavior through the process of natural selection. The most fit are also the animals that are most closely aligned with their environment. Animals employ three categories of behavior to adjust to their environment: circadian, foraging, and orientation.

Circadian

Circadian rhythms are internally controlled actions that an organism repeats daily in the same time frame. These biorhythms are responsible for waking up the entire animal kingdom on a regular basis. Humans also have a circadian wake/sleep cycle; unfortunately it is often overridden by external influences such as work, school, or other daily activities.

Ongoing research identifies a special neuron zone in the hypothalamus as the actual "clock" for animals. More importantly, this zone triggers the release of hormones that regulate the entire body. The activation of this neural zone is under the direct control of the individual's DNA. That is why some people claim to be a "morning person," while others are not. They may be responding to their genetics. The timing mechanism appears to be light sensitive, so animals adjust their daily cycles in accordance with diurnal cycles. Students who stay up late studying or get up too early to study may violate their normal timing mechanism and feel abnormally tired or different. Animals in the wild maintain circadian rhythms even down to regular bowel movements!

Bionote

To test if your natural circadian rhythm is also your current rhythm, after several normal days with no stimulants or alcohol, observe whether you wake up at your "normal" time or if you sleep in.

Foraging

Foraging is the ability of an organism to obtain energy by identifying, locating, and consuming food. Animals are often classified as generalists or specialists based on their food consumption. A generalist, such as a rat, will eat just about anything from plants to animals, living to dead. A specialist, such as a koala bear, only eats the leaves of a eucalyptus tree. Animals create a *search image* of the food type to prepare the brain for sensory clues and to minimize distracting clues.

While searching for fossilized Miocene sharks' teeth in the tidal Chesapeake Bay area of Calvert County, Maryland, researchers noted that humans who had seen a shark's tooth before the search were significantly more likely to find one before those who only had received verbal clues and instructions. Animals know what they are looking for before they hunt based on their memory, which they can retrieve as a search image to improve hunting success.

The ability to forage prevents grazing animals from eating the wrong plant. Grazers do not eat all types of vegetation. In fact, some plants are quite harmful. "Loco" weed, a type of prairie plant, was blamed for the uncharacteristic demeanor of cows and horses as they traveled overland in the covered-wagon days. The "loco weed" was new and looked like something else. Both grazer and farmer learned to tell the difference. Some foragers simply do not like the taste of some plants, but crave others. Horses will not eat horseweeds, but they love clover. Sometimes plants are edible at different times during their growth cycle. Most grazers eat grasses anytime, but rhubarb and poke are only good when young and rapidly growing. Persimmons are bitter until after the first hard freeze, and young trees are edible until the secondary growth produces woody tissue.

Orientation

By moving their body to a different location, animals orient themselves in such a way as to be in or moving toward more favorable territory. There are two types of orientation behavior:

- A *taxis* is the conscious movement in a specific direction. For instance, certain game fish, such as the striped bass, move toward a flashy lure with the intent of catching it; euglena move toward light.
- A random movement, called a *kinesis*, is generalized, nonspecific directional movement in response to a stimulus. For instance, the presence of unusual chemicals in the environment causes certain protists, such as amoeba, to stop moving. Simple organisms that possess limited ability to detect their environment exhibit most kinesis activities.

Social Behavior

Social behavior is the broadest organization of animal behaviors and always involves the interaction of individuals. They are often species specific and include behaviors that the group has adopted and transmitted over generations, which enable them to respond to each other for the good of the group. Some responses are individual, such as altruism, whereas others are whole group, such as migration. The value of these behaviors is evident in the lack of intraspecie-inflicted injury, reduced training time, and summary benefit. There are eight types of social behavior: agnostic, altruism, communication, courtship, dominance, migration, parental, and territorial.

Agnostic

Agnostic behavior is ritualized aggression that seldom ends in serious injury and is displayed in numerous other types of behavior, especially *hierarchical dominance* and *territoriality.* Rather than two individuals actually fighting, they display fight, which varies in ritual from species to species. For dogs it generally occurs when one dog out-barks, growls, or physically intimidates a competitor. Agnostic behavior also includes ritualistic behavior for the defeated contestant. This formalizes the surrender and halts further displays or the increased hostility that might end in a real fight. For the defeated dog, the ritual is a demeaning posture; ears down, submissive stature, and the characteristic tail between the legs. Agnostic behavior is a strong survival mechanism because even the winner of the real fight can receive an injury that may decrease the chances of winning the next fight and minimize future reproductive opportunities.

Altruism

Altruistic behavior normally occurs within a community where one member or members sacrifice themselves for the benefit of the remainder of the community. Most people are familiar with the altruism of honeybees. Whenever their hive is threatened, individual bees sting the invader, trying to deter destruction. Sadly for the bees, when the bee stings, the stinger normally remains in the victim; when the bee flies away or is brushed off, a portion of their posterior remains, thus eventually killing the bee. Studies on altruism indicate that it has no significant effect on the genetic variability of the colony because they are normally siblings or close relatives with limited variability to begin with. However, massive suicide missions may create a genetic bottleneck effect. For sexually reproducing animals, altruism may have an effect on genetic variability, especially if the population is small.

Altruism is also demonstrated in more advanced animals as they care and tend to others in their species. With the omnipresent threat of limited resources, helping another to survive may be detrimental to the helper. However, in a number of instances herd or pack animals will nurture and raise the offspring of a deceased mother. The offspring must be able to reprogram the imprinting instinct to accept the surrogate help or perish. In all known human cultures, a common thread emerges. Someone within the population is sustained as a healer or doctor. Most religions of the world are based on altruism as the preferred behavioral mode to avoid unnecessary conflict.

Communication

Animals use a variety of communication styles to influence the behavior of other animals. Signals and combinations of signals are designed to transmit an accurate message in a modality that can be received and correctly translated by the intended audience. Modalities include voice, posture, facial expression, scent, color, or movement. Generally the more advanced the organisms and the more complex their social organization, the greater the number of signs for a specific event so as to minimize mistakes and increase the breadth of communication. Bee communication is the classic example. Scouting worker bees returning to the hive laden with nectar perform a prescribed "waggle dance" that directs fellow workers to the source of nectar. Also, in the springtime, male birds sing their species-specific mating call, which can be heard by females far away. Frogs go birds one better: They broadcast day and night to increase their mating opportunities. Curiously, sound signals are effective 24 hours a day, whereas sight is most useful during daylight hours.

The ability to communicate effectively avoids problems, such as fights, and enhances survival opportunities, such as mating. Effective communication between individuals may be verbal or nonverbal, but always appears to connote congeniality overall. Ineffective communication may result in more destructive encounters. Animals that are herd or pack oriented rely on communication signals for survival. The slap of a beaver's tail on the water and the swift turn of a pronghorn's head signify potential danger. Animals communicate all the time—count the number of people using cell phones while they drive!

Courtship

Mating behavior is one of the most important behaviors in the animal kingdom. Successfully communicating the correct mating signals is the first step toward the continuation of the species. Most bony fish have slight coloration enhancements, which are supplemented by a *courtship* ritual of "displaying" in front of the female fish by extending all fins, curving the body, and constantly swimming in front of, or near, the female. Koi often swim side by side for hours before mating. Antarctic penguins bring tiny rocks to their potential mate to signify their intentions. Dominant males select a territory and breed with all females who enter, and fight with all males who enter (see the section "Territorial"). Male black widow spiders are careful how they approach the female during mating. One false move and the male becomes a meal, which may happen afterward anyway!

Defensive

A defensive behavior is an individual and group mechanism. A defensive group behavior is typified by the musk oxen that form a circle with the female and young inside and the horns of the male pointing outside. A defensive behavior can also be submissive, such as the defeated agnostic behavior. Animals that tend to be the smaller, less well endowed of the population tend to become more defensive as a learned behavior. The ultimate defensive behavior is practiced by the opossum, who "plays dead" whenever threatened or caught off guard by a superior force.

Dominance

A *dominance hierarchy* is also called a *pecking order*, based on a chicken phenomenon where the alpha female pecks all the other hens as a sign of dominance. The beta hen then pecks all the remaining hens except the alpha hen, and so on. Everyone pecks the last hen, and if she is removed from the population, the next-lowest hen drops to the bottom. Most vertebrates have some fashion of hierarchy, which is a species survival adaptation guaranteeing that the most fit receive the first turn at all food, shelter, and mating opportunities.

Dominance hierarchy is exhibited in both male and female arenas. The alpha female wolf will often prevent the other female wolves from breeding. Male lions typically try agnostic displays first, which minimizes combat. An understanding of dominance prevents fighting and possible injury during most of the year. However, at seasonal times, the inferior members try to overthrow superior types, usually at the onset of the annual mating season. Eventually a young male challenger will defeat an older dominant male and assume the alpha position.

Migration

The regular movement of a population from one habitat to another habitat is called *migration*. Generally migrations are seasonal as a population returns to an area of replenished resources or to spawn. To avoid weather extremes and lack of food, the Canadian goose migrates to warmer climates during the autumn months and returns to various parts of Canada for the spring in time to nest and enjoy the resource largess of a Canadian growing season. The classic example of mass migration is the seasonal movement of frail-looking Monarch butterflies from temperate areas of North America to remote mountainous locations in Central Mexico, thousands of miles away! How do they find their way?

Several hypotheses attempt to explain the geographic-positioning system and internal road maps that migrating animals seem to possess. It is thought that flying animals probably guide themselves by orienting to astronomical fixtures, such as the movement of the sun or moon, or perhaps the nonmovement of the North Star or Octans, because they can migrate above clouds or fog banks that obscure identifiable landforms. The Canadian goose travels both day and night, so they may have a combination of indicators.

Another navigation hypothesis suggests that the earth's natural magnetic field helps to orient animals. Dissections of certain animals, such as pigeons, reveal magnetite, a naturally occurring magnetic mineral, in their heads. Ocean-migrating fish, including whales, appear to follow the longer coastline route rather than taking a short cut through open water. It is hypothesized that they follow the coastline and use bottom features and terrestrial landmarks to mark their progress. Certain fish, like salmon, return to the stream of their birth by remembering the smell of the stream. It is theorized that certain migratory species have developed internal maps that guide them to exactly where they set out to end up, several thousand miles away!

Parental Behavior

Certain species, especially advanced vertebrates, mate for life or at least for a reproductive season. During that period of time, they usually share parenting responsibilities such as nest building and food finding. Birds are an interesting study. They have an interesting dichotomy: Some produce *precocial* young that are born well formed and active, needing their mother only for warmth and protection. Most duck and quail young are precocious to the extreme that they can even feed themselves! The converse is *altricial* young, which are blind, bald, and helpless. Extensive parenting is required to raise the chicks into self-reliant juveniles.

The successful production of offspring activates most advanced vertebrates to participate in some form of parenting to ensure their survival. Normally this includes the basics of food, shelter, protection, education, and a sense of belonging and imprinting identity. Ducks and other large birds often extend their wings to draw in their offspring in the event of danger or for warmth. Some pet housecats instinctively hide their litter at birth, and sometimes again if discovered during the early vulnerable days. Humans are the classic example of parental behavior. Most humans extend themselves as parents to provide an expansive childhood for their offspring as preparation for adult life. Only humans have formalized education and collateral developmental activities such as religious training, organized sports, academic teams, and mentoring services to replace a lost or unavailable parent.

Territorial

Territorial behavior happens when an animal reserves a particular location for its exclusive use. This usually prohibits certain members of the same species from occupying the same territory at the same time. Like other types, this behavior overlaps with other behaviors such as courtship and dominance. A particular territory may be claimed for a variety of reasons, including food, nesting, and mating. Typically, agnostic behavior is enough to claim a territory unless there is a shortage, and then a fight may have to settle matters. Bighorn sheep are noted for their fierce head-on collisions to establish a mating territory and dominance hierarchy. The classic study of territory involves the channel catfish. Recently, aquaculture farms have started raising catfish for profit. Their research indicates that if catfish have enough room to establish a territory, they will fight almost continually, thereby reducing profits. To remedy this, catfish young are planted so heavily that no one has any territory and they get along fine!

Usually the claimant will mark the territory to ward off competitors and prevent the arrival of unwanted same-species intruders. Most dog owners have observed the ritual spot urinating of dogs at fire plugs, trees, shrubbery, tires, just about anything that another dog might happen to sniff. Whitetail deer characteristically mark trees in their area when they wear away the bark as they rub their antlers to remove the protective velvetlike covering.

Species Interaction Behavior

Over generations, natural selection has favored those species who have adapted to their environment, both plant and animal. When two or more species appear to evolve so that they have structures and functions specifically for the other, their *coevolution* creates an advantage for both species. Four types of species interaction behavior influence the dynamics of a population and community; they are commensalism, mutualism, parasitism, and predator-prey.

Commensalism

Commensalism is a type of *symbiotic* relationship in which two species interact for the benefit of one without harming or helping the other species. Clownfish and reef shrimp are able to live within the stinging tentacles of the sea anemone without being harmed by the anemone. All other animals are stung. The fish and the shrimp receive the advantage of protection, but the anemone appears to be unaffected. Remoras are marine fish with suction discs that help them to attach to larger fish, usually sharks. They remain attached to the shark until a feeding frenzy; then they detach, feed, and

Bioterms

Symbiosis is the interrelation of two or more species inhabiting the same territory.

reattach. The shark is unaffected. Egrets and other birds that associate with cattle and other grazing animals perform the most recognized commensal behavior. As the cattle graze, they scare up insects and other food sources for these opportunistic birds. The grazers are unaffected, but the birds get an easy meal.

Mutualism

Mutualism is another symbiotic relationship in which both species benefit from their interaction. Billions of bacteria behave mutualistically in the human intestine. They break down certain indigestible foods, such as cellulose, and then produce Vitamin K for the host; they get a warm, safe place to live in return. The classic mutualistic behavior is the coevolution of pollinators and plants. Pollinators such as bees and birds receive nectar as a food source in exchange for coincidentally transferring pollen to the next flower. The pollinator receives food, and the plants are able to reproduce. An interesting mutualistic behavior is the African oxpecker, which is a bird that rides on the back of oxen and other slow-moving herbivores and pecks the ticks and other parasites off of them and consumes them for their own nutrition. Both species benefit. Some oxpeckers have gone further and have pecked the oxen until they bleed and then drink the blood for nourishment. At this point, they become parasites.

Parasitism

Parasitism is another type of species interaction where one species benefits but the other is harmed. There are two types of parasites: *endoparasites* are internal, and *ectoparasites* operate on the exterior surface. Endoparasites normally try to attach to a supply of nutrient-carrying blood. Gardeners know that certain parasite wasps attach their eggs to the invasive tomato hookworm. When the eggs hatch, the larvae eat the hookworms from the inside out. Again, they are looking for a warm food supply. Refer to Chapter 20 to marvel at the ingenious mechanisms they employ to enter and remain in a warm digestive system. Ectoparasites include ticks, mosquitoes, and leeches. Ectoparasites normally have modified structures that allow them to cling on to or lightly land on their host, usually to drill a hole through the skin to

Bionote

Vampires do exist! They live in the form of small vampire bats that inhabit tropical areas and extract blood from warm-blooded hosts, usually the lower back leg of a grazing animal. Their teeth are so sharp that the victim does not feel the tiny incision.

refuel on blood or body fluids. Microscopic analyses of lice show hooklike adaptations that prevent normal dislodging from a host.

Predator-Prey

Predator-prey interaction is an advantage to the predator and death to the prey. This interrelation probably has more effect on population genetic diversity and community social organization than any other behavior because it removes potentially reproductive members. The largest animal on Earth is the blue whale; it is a predator to one of the tiniest animals, krill.

Interestingly, as prey evolve escape mechanisms, predators evolve better capture structures and tactics as they coevolve. For instance, as warm-blooded prey became scarce in the daytime, natural selection favored snakes with the ability for nighttime capture. Snakes with heat-sensitive pits, such as rattlesnakes, are able to hunt, locate, and capture warm-blooded prey in total darkness by tracking their body heat. Antelopes and Thompson's gazelles developed incredible speed-burst capacities, which are matched by the cheetahs' speed, and the lions' camouflage and pack-hunting tactics. Predation has pressured prey to evolve new structures and strategies, which have forced predators to do the same in the endless coevolution struggle.

The Least You Need to Know

- Behavior represents the evolutionary impact on every species.
- Animals have adapted a variety of behaviors to coincide with environmental influences; some are learned, others are genetic.
- The process of learning is exhibited in a behavioral change.
- Animal behavior is not constant; some animals learn via the trial-and-error method when encountering a new situation.
- Variations in the length of childhood determines the amount of learning that an offspring receives.
- Cooperative behaviors have coevolved that both help and harm the participants.

Chapter 20

Simple Invertebrates and Animal-Like Organisms

In This Chapter

- The colonial protist theory attempts to explain the development of the first animals
- The unique features of sponges that classify them as animals
- The key characteristics of the simple invertebrates
- The advancements that led to increases in animal complexity
- Some common examples of simple invertebrates
- Parasitic animals that cause common diseases and maladies

Now that you understand how all the systems work, let's put them together and see how the various types of animals have progressed in their modification of these structures. This chapter focuses on sponges and worms.

Some of you might use an artificial sponge (or loofa) these days, but back in Roman times they used the real thing. Natural sponges work as well as most synthetic ones (made from oil) because they are both mostly hollow. The only difference is the availability of the natural sponges and their ability to last after repeated body or car washings. As a group, sponges represent the simplest animals.

Before you read the sections on worms, you may want to brace yourself. Not all of them are like the kind you want in your garden or on your fishhook. Although most are free-living and not harmful, they have ugly cousins that are parasitic on animals, including humans! Although their lifestyles are unpleasant, this section isn't. I have removed most of the cruelty and placed the worms in a realistic, yet acceptable treatment (although you might think twice about eating raw meat).

Evolutionary History: Colonial Protist Theory

The evolution from a plantlike organism to an animal-like organism was quite an evolutionary jump. The latest hypothesis states that animals evolved from plant forms in the sea. This is a reasonable hypothesis because it is believed that all life was aquatic at that time. It is theorized that nonphotosynthetic heterotrophic cells formed the basis for the first animal life. It is not clearly understood whether multicellular animals arose from individual cells joining together or a mitotic error failed to separate dividing cells. Regardless, cells became united into multicellular organisms that began the adaptive trend of cell unison and division of labor. Simple invertebrates are the first animals. They are also the most numerous: 99 percent of all animals are invertebrates.

The *colonial protist theory* is one possible explanation of how animals first formed. Because fossil evidence is nonexistent, everything is based on scientific speculation. The theory can be explained in three steps:

1. Individual cells joined together to form a hollow sphere that kept getting larger as more cells were added. The hollow sphere shape is reasonable because it provides each cell with contact to the environment and produced the protective spherical formation. As a drop of oil forms a sphere in water, the surface tension may have formed a sphere as well. If so, it marks the beginning of differentiation because the side of the cells that face the inside have a slightly different environment than the part that faces the outside. It is just as likely, however, that multicellular characteristics began as a line of identical cells, which then contorted into a sphere. The concluding evidence is still lacking.

2. As colonial protists grew in size, it is theorized that individual cells or groups of cells assumed separate functions. This would mark the beginning stages of cell interdependence, a characteristic of a multicellular organism. The next major advancement is the development of an internal layer of cells.

3. It is generally believed that the colonial group of flagellated heterotrophs invaginated, like someone stepping on a flattened soccer ball, so that the internal sides were next to each other. The central cavity that formed would have an opening to the outside when the invagination occurred. The cavity would then be surrounded by a double layer of cells. The essential description of the simplest invertebrate animal, the sponge, is a central cavity surrounded by a large layer of cells.

Interestingly, the hollow sphere and subsequent invagination are similar to the blastula and gastrula embryonic stages for all animals!

Simplest Invertebrates

Although you probably don't even realize it, you have actually been very close to a number of simple invertebrates. It's more likely, however, that you call them pests, not pets (because many of them wreak havoc on human health, food-crop production, and livestock husbandry). Known by various names throughout the world, some of the most common simple invertebrates include beautiful creatures such as sponges, sea flowers, and sand dollars as well as flukes, leeches, and eyeworms.

The simple invertebrates are a very heterogeneous assortment. The primary characteristic that unites invertebrates is the absence of a backbone. However, their body styles vary from the absence of any body symmetry and true tissues as are characteristic of sponges, to the well-defined radial or bilateral symmetry and specialized parts found in other invertebrates. Some invertebrates have no circulatory system while some molluscs and annelids have a closed system. These examples of extreme variation may be one reason why invertebrates contain the biggest number of animal species in addition to having the greatest number of individual animals alive at this moment. Similarities in body plans and their related patterns of development provide a means of classifying and establishing possible relationships among invertebrates.

Porifera: Sponges

Sponges are the simplest multicellular animal. Their phylum name, *Porifera,* says it all—composed of pores. Because all porifera are sponges, we can use their names interchangeably. Most of a sponge's body is hollow and full of pores. The openness allows water to flow into and out of the body cavity as it chooses. Sponges are little more than a group of animal cells that work together for the common good. They do not have a mouth, eyes, head, or most other structures that people associate with animals. In fact, their cells are not even organized into tissues.

Sponge Characteristics

In earlier times, biologists did not know whether to classify sponges as plants, animals, or something new. They appeared to have key evolutionary characteristics of both:

- Inhabit both fresh water and salt water (marine)
- *Sessile*, meaning they remain attached to the sea floor or other submerged object, making sponges look like plants
- Lack true roots and are held in place by "holdfasts," which are specialized cells that grow and lock into the substrate
- Lack body symmetry and cell organization at the tissue level
- Complete all of life's simplest functions: eat, eliminate waste, and reproduce
- Sponges reproduce both sexually and asexually

Sponges are passive feeders. Because they are sessile, they filter water for gaseous exchange, and small organisms and plankton for food. The water containing the foods normally enters through the many pores of the sponge and into a central digestive cavity where the foods are trapped by hairlike projections. The food is then absorbed and digested inside each cell. As the water leaves the sponge, it conveniently washes away any waste materials, including carbon dioxide. They often take advantage of tidal flow to provide a continuous source of water to bring in food and remove their wastes. Certain types of sponge appear to wave in the wind as they move back and forth in response to prevailing water currents.

Bioterms

Sessile describes any creature that attaches itself to a foundation and does not move.

Bionote

Sexual reproduction is the union of a set of gametes from two parents. The offspring therefore do not have the same genetic complement as the parents. Conversely, with asexual reproduction, the offspring are genetically identical to the parent.

Sponge Reproductive Cycle

Sponges have an interesting sex life. They are a classic example of an organism that can reproduce both sexually and asexually. Although some sponges are *hermaphroditic*, meaning they contain both male and female reproductive apparatus, most sponges have separate sexes.

The process of sexual reproduction in sponges is not much different from that of more complex animals.

In simplest terms, male sponges release a cloud of sperm into the watery environment that flow with the prevailing water currents and land on a neighboring female sponge. Obviously, this type of external fertilization is very inefficient when compared to the more common internal fertilization of more complex animals. When the sperm unites with the egg of a female sponge, the fertilized egg then develops into a larva. As the larvae mature, they are released from the female sponge and swim until they attach themselves to a suitable site where they grow into mature sponges.

Asexual reproduction in sponges generally takes two forms: budding and *fragmentation*. In budding, a tiny "bud" grows on the adult sponge. Upon maturity, the bud detaches and floats with the water currents until it is able to reattach at a suitable location. From there it grows into adulthood. In fragmentation, a piece of the sponge may be broken off by an outside force. The broken piece then attaches to a substrate and grows into an adult.

Cnidarians

In general, cnidarians are soft-bodied aquatic animals that are *radially symmetrical* and have specialized cells, such as stinging cells, that carry out specific functions. Because they do not photosynthesize, they can inhabit dark areas where green plants cannot survive, which creates the potential for expansion of additional life-forms.

Bioterms

Radial symmetry describes a body style in which the body appears to radiate out from a central point. Radially symmetrical organisms tend to have a circular body style.

Almost anyone who has gone to a warm saltwater beach is familiar with a common cnidarian, the jellyfish. These gelatinous blobs wash up on the beach and create hours of pleasure for children with sticks and a painful stinging sensation and/or a rash for unwary pedestrians or swimmers. However, there are many other types of cnidarians that are harmless, beautiful, and even helpful to the environment. Sea anemones are often brightly colored and are sometimes called underwater flowers. Among the most beneficial cnidarians

Bionote

The hard shell that corals secrete is made of calcium carbonate. This chemical compound is the one utilized by most marine animals that create their own shells. When heat, pressure, and geologic time are applied to cast-off calcium carbonate shells, they morph into the sedimentary rock limestone.

are the corals. Corals are tiny animals that secrete the telltale hard shell and sometimes, in great numbers, they create coral reefs.

Cnidarian Characteristics

Cnidarians are considered more advanced than sponges because they have …

- Radially symmetric bodies constructed with two separate cell layers arranged as tissue.
- A gastrovascular cavity that serves as a common site for digestion.
- A nerve net that primarily signals a response to specific stimuli.
- Specialized stinging cells that immobilize prey.
- Life cycles that includes two different stages.

Bionote

The stinging cells, nematocysts, are used for food capture and protection. They contain enough poison to kill or immobilize prey and are often activated by the presence of a foreign protein. They hurt when you swim into them because of the injection of the poison and because a victim almost always contacts more than 100 or more nematocysts at one encounter.

Cnidarians use their *nematocysts*, or stinging cells, to place their prey into a hopeless position. Their tentacles then move the helpless prey through the mouth into the gastrovascular cavity. Enzymes are secreted by the cells lining the gastrovascular cavity and begin the extracellular digestion of the food source, outside of the cells. When digestion is complete, the food is then absorbed by the cells. Indigestible food and other wastes are pushed back through the mouth and into the environment. Cnidarians feed on a variety of foods from very small plankton to small fish.

Cnidarian Life Cycle

Cnidarians have a creative life cycle that alternates between the common medusa, or sexual stage, and the less-seen polyp, or asexual stage. The medusa stage looks like a gelatinous blob or umbrella in the water. The male and female medusa produce sperm and egg and release them into the water environment. The egg and sperm unite to form a fertilized egg, or zygote, that grows into an embryo. The embryo matures and attaches to an object and begins to grow. Upon maturity, as a polyp it forms buds that can grow and add to the overall size of the polyp or they can detach and become either male or female immature medusa. These immature medusa are free swimming and are free to move to another location as the currents dictate. The immature medusas grow and become adults, who are then capable of producing egg and sperm, respectively.

Cnidarians are the first animal to exhibit cells organized as tissue. Individual cells have specialized functions that continue to create the division of labor and the interdependence of cells.

Platyhelminthes: Flatworms

Most people think of friendly, garden-variety earthworms when they think of worms, but this section is about their "cousins" of the phylum *Platyhelminthes*, the flatworms. They represent a vast array of worms that inhabit this planet, including the inside of humans. Certain worms have interesting life cycles that are harmless and easily ignored, but several, such as blood flukes, cause human maladies such as shistosomiasis.

Bioterms

Platyhelminthes is the phylum for flatworms, which are the simplest animals with bilateral symmetry, where each half of an organism is the mirror image of the other half.

Platyhelminthes Characteristics

As we move up the ever-increasing level of animal complexity, flatworms are the simplest animals that exhibit bilateral symmetry. They are considered more complex than cnidarians because they have the following specialized structures:

- A third or middle layer of cells organized into tissue called mesoderm, *acoelomate*
- Tissues organized into organs
- Bilateral symmetry with a distinct anterior (front) and posterior end

Flatworms are distinguished by the following key evolutionary characteristics:

- Flat bodies with no body cavity
- Digestive system includes a mouth but no anus
- Free-living and parasitic forms
- Cephalized, with a definite head
- Specialized cells that perform excretory, nervous, and reproductive functions

Flatworms are ribbonlike in their appearance. Their tapelike body structure serves as an advantage because it allows each cell to be close to the surrounding outside environment. This structural orientation allows the flatworm to easily diffuse oxygen into, and diffuse carbon dioxide out of, the cells in exchange with the environment. Their branched gastrovascular cavity extends to nearly every cell, bringing food within easy reach. Therefore, they do not require and do not have a circulatory or respiratory system.

There are two distinct types of flatworms: free-living aquatic types that inhabit streams, lakes, and oceans; and the type that are parasitic and live in warm, nutrient-rich environments, such as the inside of mammalian digestive tracts.

Free-living flatworms, such as the planaria, are common in most temperate areas and are harmless. They have several distinguishing characteristics, such as a muscular tube called a pharynx that forces food into their gastrovascular cavity. Like cnidarians, these flatworms expel metabolic and digestive waste via their mouth back into the environment.

Free-living flatworms also have a much more developed nerve net than cnidarians. Many flatworms also have one or more eyespots that are sensitive to light. These eyespots do not focus on objects (as does the human eye), but they are capable of detecting light. This information is processed in a very simple nerve grouping that also serves as the control center for their nerve net. The cephalization of the nerve net is the beginning of a brain in more advanced animals. The nerve center senses stimuli and reacts by having the flatworm move in one of two methods of locomotion. *Cilia* propel the organism in a systematic motion, by moving in unison to food and away from danger. The other mode of transportation is accomplished by simple structures that allow the flatworms to twist and turn.

Bioterms

Cilia are specialized cell projections that look and function like oars on a boat. They normally surround an organism and work in concert to move the organism to a desired location.

Platyhelminthes Life Cycle

Free-living flatworms can reproduce both sexually and asexually. Because most are hermaphroditic, containing both male and female reproductive units, they can unite with another worm and, curiously, exchange genetic material, sperm and egg, and thus both adults are fertilized. Asexual reproduction can occur when pieces of the free-living flatworm break off, or fragment, and then grow into adults.

Parasitic flatworms have much more complicated sexual reproduction life cycles. These life cycles almost always involve an intermediate host such as a snail.

Parasitic flatworms cause several human diseases. Among the most distasteful is shistosomiasis, which is caused by a blood fluke. Although they are quite small, usually less than .4 inches or 1 centimeter, the free-swimming larvae of this blood fluke bore into the skin until they are able to enter the bloodstream. Once in the bloodstream, they travel throughout the body and eventually attach in the intestine, where they grow into adults. Common in tropical areas of Africa and Southeast Asia, the resulting disease weakens the victim, often human, to the point of death.

Another popular parasitic flatworm is the tapeworm, often called man's best friend because it goes where you go and eats what you eat. Tapeworms are contracted most often by eating the improperly cooked meat of an infected animal. Once inside the host, the larvae circulate in the blood and attach to the intestinal wall. From there, it simply remains attached to a continual source of life-sustaining nutrients. As a prudent parasite, the tapeworm seldom kills its host, instead preferring the symbiotic relationship.

Bionote

Human tapeworms have been measured at a length of over 59 feet or 18 meters!

The bilateral symmetry and cephalization introduced by flatworms are an advancement because they are the first to be able to move headfirst into their environment. This is an advantage because they can orient their specialized sensors such as eyespots to help guide their way. This type of body organization also contributes to greater division of labor.

Nematoda: Roundworms

Nematodes, or roundworms, are extremely adaptable and are found almost everywhere living things are found. They have a tough cuticle body covering that serves a protective function. They are also structured with an enlarged head area compared to the thin, tapered tail. They are slightly more advanced than flatworms. Common examples include hookworms, and trichinella, the worm that causes trichinosis.

Nematoda Characteristics

Nematodes' key identifiable evolutionary characteristics include several anatomical advancements:

- Beginning of a true body cavity.
- Only animals with a pseudocoelom.

- Simplest organism to have a one-way digestive system that includes both a mouth and anus (alimentary canal).
- Composed of three tissue layers.
- Extremely abundant.
- Similar body style that is round and threadlike.
- Most are free-living, but some are parasitic.

The body cavity is a major advancement because it aids in movement, protects internal organs, and promotes circulation of fluids.

A pseudocoelom is a fluid-filled cavity located between the digestive gut and the body wall in roundworms. It is in contact with both the muscle layer of the body wall and the lining of the digestive tract. The pseudocoelom permits the slow diffusion of nutrients into the gut, which in turn limits their overall size. Fluids within the pseudocoelom exchange oxygen and carbon dioxide as well as distribute nutrients to all cells. This serves the function of a circulatory system and a digestive system. All animals more advanced than nematodes will have a true body cavity.

Nematode Life Cycle

Roundworms do not have a romantic reproductive lifestyle, although they do have separate sexes, both male and female. After mating the female deposits the fertilized eggs in an environment that is suitable for roundworms—just about anywhere. The microscopic eggs hatch and grow into adults.

Certain parasitic roundworms have a more complex life cycle involving several hosts. Roundworms are so prevalent that they infest almost everything, both plant and animal. Probably the most common parasitic roundworm is the pinworm, *enterobius*. Pinworms affect slightly less than 20 percent of adults and almost 30 percent of children worldwide. As prudent predators, pinworms are an annoyance, but do not normally create a fatal disease. The adults resemble white thread, live and mate in the lower large intestine, and are quite small, less than 2 inches or 5 centimeters. They are quite active at night while the host sleeps. The female slides out of the intestine, lays fertilized eggs on the outside of the host's body, usually around the anus, and then retreats back into the warmth and shelter of her intestinal home. When the unsuspecting host scratches during sleep or at other times, the eggs are picked up on the hands, often ingested, and then migrate to the intestine to begin their next life cycle. The eggs can be spread to other locations and other animals. When an unknowing person with microscopic eggs on his or her hands touches you or something you are about to touch, the malady may be spread.

Another common worm is the *ascaris,* which is an intestinal worm that infects humans, dogs, cats, cows—almost everything with a warm digestive tract. Usually the infestation spreads when a healthy organism ingests, either by eating or licking, the fertilized eggs of the parasitic roundworms. The eggs hatch and attach to the blood-fed, nutrient-enriched walls of the intestine and grow to maturity. When a pet owner has a pet "wormed," it is to rid the pet of *ascaris.* Because the infestation spreads so easily, for livestock owners *ascaris* can be very expensive to cure, control, and prevent.

Another interesting disease is caused by a species of *Ascaris lumbricoides.* When immature, the *lumbricoides* worm often drills through the lung tissue and may cause pneumonia before attaching to the intestine as it matures into an adult. Another more pronounced malady is also one a lot of people have seen in gruesome pictures: elephantiasis (not elephantitis). The circulatory parasitic roundworm *Wucherena bancrofti* creates this malady by growing and clogging the lymphatic system, which prevents the normal flow of lymph fluid. This creates a bottleneck and the resulting backup of fluid is seen as a painful and skin-thickening gross, often grotesque, swelling of limbs, usually the legs. In the past, pictures were taken for medicinal reasons and for people to gaze at in wonderment.

Bionote

Heartworms are roundworms that also infect dogs, cats, and other pets and use the mosquito as an intermediate host. In certain temperate areas, pets must stay on heartworm medicine all year to prevent an infection.

Bionote

Now you know why your mother told you to wash your hands before eating and never to eat raw meat.

Mollusca

The word *mollusc* comes from the Latin word meaning "soft." Considering the phylum Mollusca (a.k.a. Molluska) often come in hard shells, this description seems misplaced. Actually, the soft part of the mollusc is inside the shell, plus, several species of molluscs have no shells.

A number of seafood lovers are also mollusc lovers. Why is that? Because oysters, clams, scallops, and squid are prime examples of molluscs. Gardeners also know two other types of mollusc: snails and slugs. The phylum Mollusca is the second largest phyla on Earth (behind the Arthropod phyla).

Mollusca Characteristics

Although some molluscs, such as the clam and oyster, have a shell, others, such as the squid and octopus, do not. What are the key identifying characteristics then? The defining key evolutionary characteristics that represent a leap in animal complexity are the presence of several specialized structures:

- Coelom
- Closed circulatory system
- Excretory organs

The coelom is a body cavity that is completely lined with mesoderm tissue. The mesoderm suspends and separates the digestive system from the movement of the remainder of the body. Likewise, the flexible nature of the coelom permits an organism to twist and turn its body, which enables greater movement. The embryonic coelom permits physical contact between the endoderm and mesoderm that leads to organ development. A true coelom also protects internal organs from injury. For instance, sharp blows and heavy hits are modified by the insulating effects of the coelom. It also protects areas beyond the protective influence of the skeleton.

Another advancement in efficiency is the closed circulatory system. A closed system is less random and much more direct than an open system and therefore delivers blood more efficiently to the entire body.

Finally, the advancement in the excretory organs, called *nephridia*, further distinguishes the molluscs from the worms. The nephridia are structures that filter cellular products from the coelom and remove them from the organism (see Chapter 14). The *radula* is an interesting adaptation for scraping food off of a surface. It is a rasping tonguelike structure with rows of teeth like a carpenter's file. In parasites it has been modified into a pointed, needlelike structure that stabs prey and injects venom.

Bionote

Certain oysters and sea slugs are able to switch sexes. Sometimes they are male, sometimes they are female.

Mollusca Reproduction

Molluscs do not demonstrate any new advancements in reproduction. Although certain species are hermaphroditic, they are generally either male or female.

Molluscs that have two external shells and most snails release both sperm and egg into the aquatic environment. The sheer number released ensures a free-swimming trocophore will develop from a fertilized egg. For the remaining molluscs, fertilization is internal. Interestingly, certain hermaphroditic species pair with a partner and exchange gametes. So both female sections are fertilized by both male sections simultaneously!

Molluscs have a varied role in the ecosystem. Molluscs are filter feeders that clarify water and are important in the removal of detritus (decaying plants and animals). They are an integral part of the food web, serving as a predator, prey, scavenger, parasite, and parasite host. Slugs and snails eat garden plants, and shipworms, the "termites of the sea," drill holes in the hulls of wooden boats. Molluscs are much more plentiful and diverse than nematodes.

Annelida

Annelids are the most recognizable worms for most people. They are long and cylindrical, with a segmented body. The common garden earthworm is a great example. Annelids are found almost everywhere, both in marine and terrestrial territories. However, except for the earthworm, they are seldom seen by people.

Segmented worms are an evolutionary advancement, most notably because of their body style, which appears to be made in segments (hence the name, segmented worms). Annelids have several key evolutionary characteristics:

- Sensory organs
- Segmented bodies
- Complex nervous system
- Closed circulatory system
- Respiration through their skin

Many annelids have well-developed nervous systems and sensory organs that allow them to be sensitive to their environment. Free-living marine annelids have the most well-developed sensory organs. Some marine annelids have sensory tentacles, chemical receptors, and two (or more) pairs of eyes that serve as light detectors. In addition, they usually also have specialized sensory organs in their epidermis which respond to vibrations, certain chemicals, and light. The common earthworm relies heavily on the sensory cells in its epidermis to detect its environment.

The Annelids contain roughly 15,000 species which are segmented and bilaterally symmetrical worms. Their body segmentation is their most recognizable feature. Interestingly, each segment can operate somewhat independently of the neighboring segments. Because of this, the segmented worms demonstrate a degree of tissue specialization. As an example, the sucker-type mouthparts found in leeches are specialized to attach and suck body fluids or blood from a victim.

Some segments are modified for specific functions, such as reproduction and sensory reception. An anterior segment is specialized as a cerebral ganglion, or proto-brain, which is connected to a ventral nerve cord. Interestingly, several Annelid species are quite active with well-developed nervous systems. Their brain is located anterior to the top of their digestive gut, with large nerves passing on both sides of the gut connecting to ganglia located below the gut. From these ganglia a nerve cord runs on the ventral side for the entire length of the worm providing nervous connections throughout the body. Each segment has its own nerve connections into the central system that carry messages to and from the sense organs and coordinate the movement of muscles.

Most Annelids have a closed circulatory system that consists of two main blood vessels that run on the dorsal and ventral sides for the length of the body. Smaller vessels called ring vessels connect the dorsal and ventral vessels in each body segment, providing a supply of blood for the internal organs. In earthworms and several other Annelids, anterior ring vessels are enlarged and reinforced with muscle tissue to contract rhythmically to help pump blood through the system.

Bionote

Leeches are making a comeback as a medicinal tool! In the past, leeches were sometimes applied by the learned physician of the day to suck out "bad blood" from a patient. Unfortunately for the victim, this often made matters worse. Presently, the blood-sucking leeches *Hirudo medicinalis* have been used by learned physicians of the day to liquefy blood clots and remove blood from traumatized tissue and extensively bruised areas.

Although some aquatic annelids breathe through gills, an interesting characteristic of many annelids is the ability to breathe through their skin. In order for the oxygen to diffuse into the body and carbon dioxide out of the body, the skin must be kept moist. In fact, if they dry out, they usually die. To prevent this from happening, many terrestrial annelids, including earthworms, secrete a protective mucus coating called a cuticle to help keep the skin moist. The gaseous exchange does not occur unless the membrane is moist.

Annelids have a biotic role that is often never seen by humans. Their larvae constitute an important aquatic food source. Earthworms aerate the soil and in so doing mix important nutrients such as nitrogen for reuse by plants. Modern medicine may revive the use of leeches in certain medical treatments! The common earthworm is also popular with gardeners because earthworms aerate the soil and leave nutrient-rich castings wherever they travel. Earthworms also have a brain, a nerve cord, a circulatory system that branches to each body segment, and an extensive digestive gut (see Chapters 14 and 15).

Ever wonder why there are so many earthworms on your concrete sidewalk or driveway after an overnight rain in the spring? The latest earthworm research indicates that these are the ideal conditions for worm mating. Once mated, the worms are supposed to crawl back into their holes.

The Least You Need to Know

- Porifera is the phylum that includes all of the sponges and is also the first multicellular phylum.
- Cnidaria are aquatic animals that exhibit radially symmetric body styles and nematocysts or stinging cells.
- Platyhelminthes contain the flatworms, which are the simplest animal to exhibit a bilaterally symmetric body style.
- Nematoda contains the roundworms, which are among the simplest animals to have a digestive system with two openings; this allows the unidirectional flow of food from the mouth to the anus.
- Mollusca is the first phylum to possess both a coelom and a circulatory system. The use of organs and organ systems increases the specialization and efficiency of that organism.
- Annelids are the first to exhibit segmented body parts that begin the advancement of body-segment specialization.

Chapter 21

Advanced Invertebrates

In This Chapter

- Unique and abundant arthropods
- Book lungs increase respiration while decreasing evaporation
- Jointed appendage modification
- The characteristics of spiders, insects, and crustaceans
- Structural advantages for a complete or incomplete metamorphosis
- The significant structural advancement of mandibles

This chapter begins the transition from invertebrates to vertebrates. Advancements made in simpler organisms are discussed here and further developed in the next chapter.

We begin with arthropods, which are virtually everywhere and in almost every environment that living things can occupy. People love to eat them (lobster, for instance), and people love to hate them (ticks, for example). Whatever the occasion, there is an arthropod that fits the theme. We will examine two classes of arthropods that collectively exemplify everything that arthropods stand, or fly, or swim for.

Echinoderms are easy to cover because they don't move as fast and there aren't as many of them. Invertebrate chordates are even easier. Most people will live their entire life and never see a living invertebrate chordate. They are a relatively unknown species except that they are the transition point between the most complex invertebrate and the simplest vertebrate. They set the stage for all vertebrate characteristics. Arthropods are found almost everywhere.

Arthropods

Arthropods are the most diverse animal phyla, with almost one million species identified with the promise of more to come. Most people are familiar with common arthropods such as spiders, ladybugs, millipedes, and maybe even scorpions, but when is the last time you saw a mite, pill bug, centipede, or horseshoe crab? The key evolutionary characteristics for arthropods include several new advancements that separate them from simple invertebrates:

- Jointed appendages that are adapted for specialized functions
- Molting the chitin-sheathed exoskeleton as it grows
- Malpighian tubules in several species for excretion
- Book lungs for increased respiration

The term *arthropod* come from the Greek word *arthros*, meaning "jointed," and *podes*, meaning "feet." Arthropods do not really have jointed feet, but they do have jointed appendages, which is an identifying characteristic and advancement in complexity. Jointed appendages not only permit walking; the appendages are modified for jumping, holding, and sensing the environment.

The body segmentation initiated by the annelids (in Chapter 20) is expanded upon by the arthropods. For instance, although caterpillars are segmented like annelids, the resulting butterfly has three characteristic body segments typical of arthropods: the head, thorax, and abdomen. Body segmentation allows for organ-system specialization, which is an advancement in efficiency. Fusion of individual body segments into these three parts also creates a tough exoskeleton that is used for protection and muscle attachment.

The exoskeleton is made of chitin, a chemical compound composed of a polysaccharide and a protein. Because the exoskeleton does not grow, the animals must shed their exoskeleton, or molt, as they outgrow it. Molting can occur at any time but only occurs as a result of growth. During the molting process, the organism usually backs out of the shell as enzymes separate tissue from shell. This is also the most vulnerable time for the organism. Once out of the old shell, the soft body parts are an easy meal for predators. Normally the new shell forms within one hour, completing the entire process within a six-hour period for most species.

Three common examples of arthropods have been selected as representatives to illustrate advancements made by arthropods as the life story of spiders, insects, and crustaceans unfolds. Each is described in a section that follows.

Arachnida: Spiders

Spiders are an interesting study of a typical arthropod. Most of the more than 35,000 species of spiders are harmless and even beneficial to humans. Most live on land, but a few, such as the water strider, live in freshwater. However, it is suspected that spiders were the first arthropod to move from an aquatic habitat and colonize on land. They demonstrate several of the key evolutionary characteristics exhibited by arthropods that make them unique:

- *Malpighian tubules*, which help maintain their water balance.
- Elimination of indigestible foods and other wastes without upsetting their internal water balance.
- *Book lungs*, which increases the surface area for respiration without increasing evaporation.
- The production of a strong, flexible protein that is silklike. Often they spin elaborate webs or use the thread as a rope ladder or swing for transportation.

Bioterms

A **book lung** is a folded air sac that looks like the pages within a book. These numerous folds provide extra surface area for greater gaseous exchange and therefore greater efficiency. They conserve water because the air is drawn into a moist membrane that allows the oxygen to readily diffuse into the blood. Because the book lungs are internal, the moisture does not evaporate to the environment.

A major problem for animals as they moved from an aquatic to terrestrial habitat was water conservation. While in a water environment, it was never a problem; once on land, however, it became a limiting factor. The advent of an excretory and respiratory system that conserved and maintained an acceptable water balance was an advancement to land colonization.

Many spiders also use book lungs to breathe. Book lungs are connected to the outside environment. This advancement promotes a more efficient gaseous exchange. Book lungs allow a greater range of movement by increasing the amount of oxygen that is available for respiration and removing the waste carbon dioxide more efficiently.

Spiders are predators, and some spiders can even produce poison. The poison they produce works well to immobilize captured prey. Although they are nature's control for a burdensome insect population, several species have been known to catch fish, frogs, and even small birds. As a class, they have developed several interesting techniques for capturing prey. For instance, some actively chase their prey, others ambush their victims, whereas others hide beneath trap doors, and of course the familiar, ever-present spider web entrapment model.

Usually the spider dispatches the victim with a bite from its fangs, sometimes injecting poison. Some also wrap their victims in a silk body bag for increased containment and later consumption. After the spider has secured the prey, it uses a pair of modified jointed appendages, called *chelicerae*, to inject paralyzing venom into the prey. Then enzymes produced in the spider's mouth break down and liquefy the tissue for easy ingestion. The liquefied contents are drawn into the spider's digestive tract by the combination of their esophagus and a specially adapted pumping stomach that brings the liquid nutrition into the spider, much like a human devours a milkshake.

Bionote

Different species of spiders spin their webs in different trapping locations to entrap specific victims. Some create vertical webs to trap flying prey. Others create horizontal webs to trap falling or wind-blown food. The most creative spider-developed snares are developed as elaborate underground traps for unwary victims.

The American black widow spider and the brown recluse spider produce a venom that is dangerous and sometimes fatal to humans. The black widow is common in temperate areas, and the venom from the bite of a female acts on the victim's nervous system. These spiders can be identified by a red-orange hourglass design prominently displayed on their abdomen. Brown recluse spiders like to hide in dark places, such as shoes and basements, but thrive in almost any temperate area, although they may not be as widespread as at once thought. Being reclusive, they prefer to stay hidden, making them even more dangerous to the unsuspecting victim. To make matters worse, they inject a venom that causes extensive internal tissue damage for a wide area around the initial bite site. They have an easily identifiable violin shape design on their upper body segment.

Another often-deadly activity regarding spiders is mating. Because spiders are generally solitary creatures, it is important for the prospective mate to understand that the suitor is not a convenient meal. In the spider world, elaborate rituals prevent the male, who is generally the smaller of the two, from being consumed by the female. Female black widows sometimes eat the male after fertilization anyway, hence the name black widow.

Insects

The class *Insecta* is also an important and representative example of arthropods. They have shown great staying power in the battle for survival. Insects are known to have predated dinosaurs and may have thrived for more than 300 million years. Currently it is estimated that insects outnumber humans by at least 200 million to 1. In fact, insects are so prevalent that there are more than 900,000 named and recorded species, and they account for almost three fourths of all the species on Earth!

Bionote

Almost one-half of the named and recorded insects are beetles, roughly totally 300,000.

In the same way that an up-close study of spiders illuminated additional biological principles, a similar study of insect key evolutionary characteristics helps define arthropods as a phylum:

- Body segments are fused into three distinct body regions: the head, thorax, and abdomen.
- Three sets of legs, one set of antennae, and a set of complex mouthparts.
- Modified antennae that serve many functions.
- Mandibles (movable jaws) that allow greater chewing angles.
- Legs that are highly adapted to function.
- Enzyme specificity.
- Complete and incomplete metamorphosis.

Most adult insects also have two pairs of wings for conflict or flight. They are made of chitin for structural support. Examine the following illustration *Typical insect*.

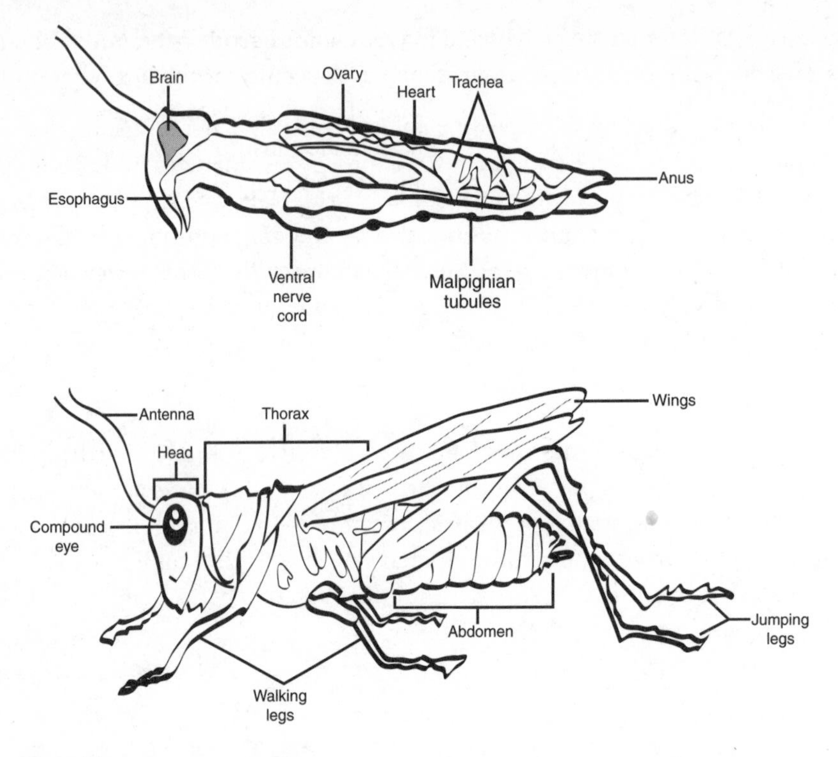

Typical insect.

Probably the most identifying feature in insects is the adaptation made to create specialized structures from the jointed appendages. Insects are the first to exhibit appendages specially adapted to sense their environment. The antennae are attached directly to the head of the insect, so they move their head in all directions to angle their antennae for maximum reception, much like owners of televisions with rabbit-ear antennae.

Interestingly, another advancement is demonstrated in insect legs. These jointed appendages are specialized for jumping (fleas), or walking (ants), or hanging on to hair (lice).

Bionote

Insects, birds, and bats are the only animals that can fly unassisted. Insects are the only invertebrate capable of flight. They were also the first animal to fly.

Simple insect species have mouthparts that are useful for holding and eating food. More complex insects have further modified their mouthparts for sucking, piercing, sponging, and for reaching nectar at the bottom of long, tubular flowers. Insects also have movable jaws, called mandibles, that allow them to eat and bite a great variety of foodstuffs from plants (crops) to animals (you). Insects have specialized mouthparts that are formed to fit their function and the accompanying enzyme specificity. Like their mouthparts, insect enzymes are specific for the type of food they prefer. In other words, a grasshopper has a mouthpart and accompanying enzymes that are designed to eat your flowers or crops, and a female mosquito has a mouthpart and enzymes especially designed to pierce skin, suck, and digest your blood.

Collection and excretion of wastes is handled by the *Malpighian tubules.* Interestingly, in certain insects, these tubules remove so much water from the excrement that it is almost crystalline. (Refer to Chapter 14 for a detailed description.) This conservation of water is very important for regulating the overall water retention and balance in the organism.

Insects are about as romantic as dirt. They do have separate sexes, and the male does deposit sperm in the receptacle of the female. The only interesting part is that the sperm is stored in the female's seminal receptacle until the eggs are presented for fertilization by the ovaries. At that time, fertilization happens quickly, and when ready, the female deposits them in a suitable location, often underground. Once fertilized eggs grow and hatch, they become interesting again. Depending on the species, the offspring may undergo either a complete or incomplete metamorphosis as they grow to adulthood. See the illustration *Incomplete and complete metamorphosis.*

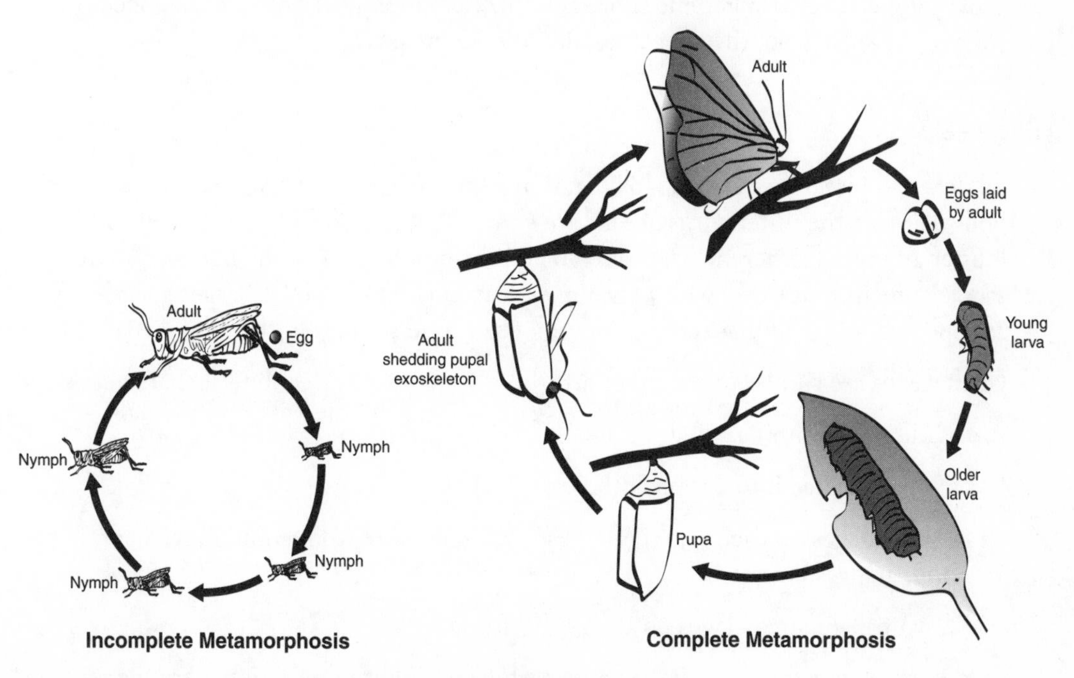

Incomplete and complete metamorphosis.

During *complete metamorphosis*, the egg hatches into a *larva*. The larva is often called a caterpillar and is obviously wormlike and does not look like the adults. During this phase, the insect can be very destructive, especially to your plant friends, because of its ravenous appetite. Its main function at this stage of development is to eat and grow ASAP. After gorging during the larva stage, the larva matures into the *pupa*, which is characterized as a developing stage in which the insect is encased in a protective exoskeleton covering.

Bionote

In moths, the exoskeleton is called a *cocoon;* in butterflies, a *chrysalis*.

Although the pupa phase does not look like the adult, it is in the pupa stage that the insect changes into its adult form. It has been said in some circles that the pupa stage is the location where a worm crawls in and a butterfly flies out. The adult emerges from the exoskeleton and mates with another insect to begin the cycle again.

Incomplete metamorphosis skips a few stages but still proceeds from a fertilized egg to an adult. The only stage between the egg and adult is called the *nymph*. A nymph generally looks like a small adult but is incapable of reproduction. As the nymph grows, it molts, eats, grows, molts, and eats, and so on. Nymphs may molt several times as they grow and become stronger and more adultlike. Eventually, if not eaten, stepped on, sprayed, or splattered, they become fully reproductive adults.

Crustacea

Spiders and insects are arthropods that dominate the land, but crustaceans are arthropods that rule the water. Crustaceans are generally well known because they include lobster and many different types of crab and shrimp. They also include myriad microscopic animals that travel with prevailing ocean currents as well as larger species such as copepods and krill that serve as an important food sources in the food web.

Crustaceans are similar to the other arthropods, with six key evolution characteristics:

- Gills for breathing
- Body divested into a cephalothorax and abdomen
- Appendages on each segment and specialized for reproduction, movement, feeding, and respiration
- Two pair of appendages on their head for feelers
- Exoskeleton that contains calcium carbonate (lime) for additional hardness
- Larval form called a *naupilus*

The lobster is a typical crustacean that has several defining characteristics that are exhibited in most crustaceans. Refer to the illustration *Typical lobster anatomy* to locate the various specialized pair of appendages attached to each body segment. Each pair has a special function such as the antennae and antennuoles that protrude from the head area and are used as sensors. Lobsters also sense vibrations and certain chemicals in the water with hundreds of small specialized threadlike cells concentrated on the antennae.

Bioterms

A **naupilus** is the free-swimming larval form that is structured dramatically different from the parent, having three pairs of appendages and a single eye located near the terminal end.

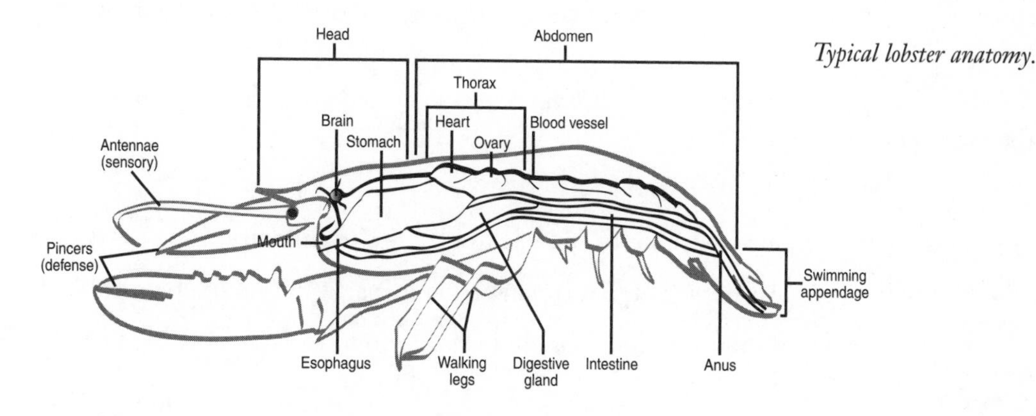

Typical lobster anatomy.

Their jointed appendages are highly modified. The *mandible* is the modified movable mouthpiece that allows for chewing while handling the food with two pairs of macillac and three pairs of macillipeds that present the food for the mandible.

Chelipids are the pincers, and the next four pairs are walking legs. There are five abdominal segments, each of which has a small pair of legs, called swimmerets, that functions to create water currents for sperm distribution and egg housing during reproduction. Finally, the oversized uropod is the posterior appendage modified for movement. In several species, the appendages are sheathed and separate from the exoskeleton for extra protection.

In digestion, food passes the mandibles, the mouth, and esophagus before entering the stomach. Interestingly, the stomach has modified teethlike tissues made of calcium carbonate and chitin to mechanically grind the food, which is then mixed with digestive enzymes secreted by the digestive gland located behind the stomach. The resulting soup passes into the intestine for absorption of water and nutrients and the completion of digestion. Foodstuffs that cannot be broken down by digestion are then eliminated via the anus. This complete digestive process is similar to more complex vertebrates, including humans (see Chapter 14).

At one time it was thought that arthropods evolved from segmented annelids. Fossilized evidence from approximately 500 to 600 million years ago, during the Cambrian period, appear to bridge and link the two together. However, DNA analyses indicate that they did not come from a common ancestor! So somewhere before annelids, the arthropods adapted their own modifications.

Echinoderms

The next move up the increasing animal complexity tree is the phylum Echinodermata—the echinoderms. *Echinoderm* is probably not a household name; however, sand dollars, brittle stars, and starfish are adornments a lot of beach tourists pick up in the local shops as souvenirs of their vacation. All echinoderms are marine animals that are often predators.

The key evolutionary characteristics of echinoderms are somewhat different from the arthropods:

- Deuterostome cell organization
- Juvenile bilateral body symmetry and adult radial body symmetry
- Centralized circular nerve net that reaches all body parts, but no brain or head
- Circulation and respiration functions are in the coelom
- Endoskeleton
- Hydrostatic water vascular system

A deuterostome is advancement beyond the protostomes. Minimizing the embryology, the developing embryonic cells of the deuterostomes, if separated, can each grow into a normal mature adult. This is not the case for protostomes; when the first cell is created, the fate of all future cells is decided.

An identifying feature for the echinoderms is the development of an endoskeleton. They are the first phylum to have an internal skeleton. Echinoderms are interesting in another way: The body style of the larvae is bilaterally symmetrical and undergoes a metamorphosis as they mature into the more recognizable adult form, which is radially symmetrical.

Although they have no brain or even a head, they are capable of controlled and coordinated movement because of an interconnected ring of nerves that branch throughout the body. (Refer to Chapter 15 for a complete explanation of the nervous system.)

The only place to find an echinoderm is in a marine environment. They have no true excretory system to maintain water balance, so they let the saltwater environment do that for them. Echinoderms do have a one-of-a-kind hydrostatic water vascular system. These water-filled canals that extend throughout the underside of the organism allow the animal to move, feed, and regulate gas exchange by adjusting the amount of water pressure in one or more of the canals. Starfish use their hydrostatic water vascular system to surround and eventually open clams and oysters, much to the chagrin of commercial fishermen. Before knowing better, watermen would angrily tear apart starfish and throw them back into the water for burial. The only problem with this scenario is that the starfish, and several other echinoderms, have the ability to asexually regenerate their entire body from a body fragment, in a process called *fragmentation.*

Invertebrate Chordates

Most people will never find an invertebrate chordate, so you might wonder why we spend time examining their lifestyle. Invertebrate chordates are the transition between invertebrates and vertebrates. Their key evolutionary characteristics are very similar in certain respects to echinoderms. They offer several very unique advancements that are worth remembering:

- Notochord
- Hollow dorsal nerve cord
- Pharyngeal slits
- Early gills
- Postanal tail

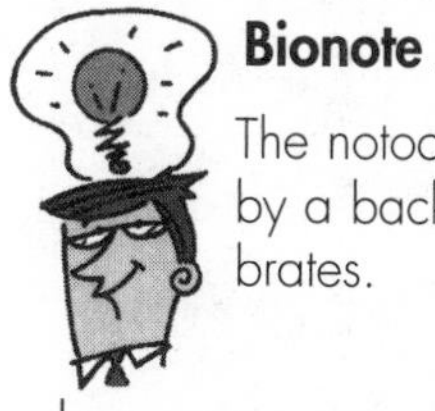

Bionote

The notochord is replaced by a backbone in vertebrates.

A notochord is a stiff, rodlike structure made of tissue that develops along the back, or dorsal side, of the embryo that extends through at least part if not most of the body. Some chordates keep the notochord throughout life, while most exhibit it only during the embryonic period. Muscles that connect the body to the notochord allow the invertebrate chordates to swing their bodies side to side, enabling them to swim.

The hollow dorsal nerve cord connects the nervous system, allowing nervous coordination throughout the body. In most invertebrates, the nervous system joins to the nerve cord at regular intervals and connects with the internal organs, muscles, and sensory organs. Normally it is found located near the notochord.

Pharyngeal slits are openings in the wall of the throat (pharynx). The pharynx is a muscular tube that connects the mouth to the digestive tract and windpipe. The purpose of the pharyngeal slits is to connect the pharynx more directly to the outside environment to increase the efficiency of gaseous exchange. This connection to the environment allows for greater exchange of oxygen into the body as well as carbon dioxide out of the body.

Early gills are often found in the pharyngeal slits. They work best when the invertebrate chordates force water past the pharyngeal slits to allow access to new environmental substrates for the gaseous exchange.

The postanal tail creates an advantage because having a tail south of the anus allows for independent mobility of the tail. Because of this, they can use the postanal tail as a means of propulsion in the water.

Interestingly, all the key evolutionary characteristics of the invertebrate chordates are always present in all chordates. They are not always visible, though. In some cases, they appear only in the embryonic stages. Because of these common characteristics, invertebrate chordates are often described as the connection or bridge between the invertebrate animals and the vertebrate animals. A common embryologic history indicates that they are both deuterostomes with an endoskeleton. Invertebrate chordates are interesting because they reveal that it is likely that vertebrate and invertebrate chordates evolved from a common ancestor. In the next chapter, the concept of similar evolution is continued, as these same characteristics are exhibited by the vertebrate chordates as we move up the ladder of complexity.

Bionote

The early gills and pharyngeal slits are the precursor for gills in more advanced fish.

The Least You Need to Know

- Arthropods are the most numerous phylum, and they have jointed appendages as a result of body-segment specialization.
- Spiders spin silk from a specialized structure in their abdomen to create entrapments for food, safety lines, transportation, and nesting.
- Insects have three main body parts: head, thorax, abdomen; and they have three sets of legs. They exhibit either complete or incomplete metamorphosis, depending on their species.
- Crustaceans have shells (made with calcium carbonate), gills, and two pair of specialized appendages on their head for sensors.
- Echinoderms are marine animals that have a water vascular system that opens to the outside environment.
- Invertebrate chordates are the transition between invertebrate and vertebrate animals.

Chapter 22

Vertebrates

In This Chapter

- Characteristics that separate vertebrates from the remainder of the animal kingdom
- The complexity in vertebrate structure and function as they become more efficient
- How man is structurally related to the other vertebrates
- Recognizing a hominid by structure
- How birds fly and why you can't
- Embryonic comparisons and ancestral linkages

At last, we have reached the home turf. This is the chapter where humans come into play. As we work through some of the simpler vertebrates on our way to a study of you, notice the increasing complexity of the organs. Also notice that they are still based on simple processes such as osmosis and diffusion.

The advanced vertebrates have structures that do not vary greatly from what you are using. Some are very close to identical. In fact, you might even remember hearing about animal organs being transplanted into human bodies.

Archaeological evidence presents a rather complete story of proto-humans and early humans. The latest data on that trail are presented in a chronological context. It is interesting to note that man stood erect before his brain fully developed into its present state. It also appears as if early-man types were a lot like us; check it out for yourself.

Chordates

The transition between invertebrates and vertebrates occurs in the chordate group. They have several characteristics of the invertebrates while exhibiting the advanced tendencies of the vertebrates.

Note that the vertebrate chordate characteristics were also mentioned in the previous chapter as we discussed the key evolutionary characteristics of the invertebrate chordates.

At one time in their development, all chordates exhibit the following key evolutionary characteristics:

- Notochord.
- Hollow dorsal nerve cord.
- Pharyngeal slits.
- Most have two sets of paired appendages, a closed circulatory system, a ventral heart, and lungs or gills for breathing.
- Most advanced chordates, including humans, replace the notochord with a backbone or vertebral column.

The notochord is a stiff, rodlike structure made of tissue that develops along the back, or dorsal side, of the embryo.

The hollow dorsal nerve cord connects the nervous system, allowing nervous coordination throughout the body.

Pharyngeal slits are openings in the wall of the throat (pharynx). The pharynx is a muscular tube that connects the mouth to the digestive tract and windpipe. The purpose of the pharyngeal slits is to connect the pharynx more directly to the outside environment to increase the efficiency of gaseous exchange. This connection to the environment allows for greater exchange of oxygen into the body as well as carbon dioxide out of the body.

Vertebrates are not the largest or most abundant group of animals. In fact, we recognize only about 45,000 species of vertebrates. Fossil evidence of the first vertebrate, a jawless fish, appeared to be about 550 million years old. Fossil records indicate that the earliest fishes were both vertebrates and chordates. As such, fish exhibit all of the key evolutionary characteristics of the chordates plus several characteristics that have been modified for specialized purposes.

Fish

Fishes are aquatic vertebrates that contain species that are considered to be the most primitive-living vertebrates. They are thought to be the first vertebrate that developed and have characteristics that might have existed in the earliest ancestors:

- Jaws (except for the jawless fish, covered in the next section), which are limited to small particles or fluids that can be sucked in.
- Paired pectoral (anterior) and pelvic (posterior) fins attached to cartilage or bone and are used for propulsion in the water.
- Muscle-skeletal interaction. This provides greater range of body movements.

Today, fish are the largest group of vertebrates in total numbers and numbers of different species. Water covers three quarters of the earth, and some type of fish lives in virtually any type of water.

Jawless Fish

Jawless fish do not really look like what most people think of when they conjure the image of a fish. In fact, they actually look more like large roundworms. Jawless fish have several key evolutionary characteristics:

- Modified mouthparts replace the need for functional jaws.
- Smooth skin lacking scales or plates.
- Internal cartilaginous skeleton. This means that their skeleton is inside of the body and made of cartilage. This characteristic allows them to undergo an interesting phenomenon: When confronted by a predator, they are able to fold their body in a knot and appear too big to fight, too big to eat, or too big to carry away.

Bionote

Most "eel-skin" products on the market today, such as wallets, are actually made from the eel-looking hagfish, one of the few remaining jawless fish.

- Unpaired fins. This means that perhaps they only have a single fin on the top of their body to aid in movement.

The lamprey is a good example of a jawless fish and is a freshwater jawless fish that escaped extinction. Most lampreys are parasitic and use modified mouth parts to attach to their prey, usually a fish. While attached to the side of the fish, the lamprey uses its specialized mouth to grind a hole in the side of the fish and then gorge on the blood and body fluids. When satiated, it disengages and the host usually dies.

Although some lampreys visit saltwater habitats, they always reproduce in freshwater, often in small streams with gravelly beds. Typical of external fertilization, the female clears a depression in the gravel and deposits eggs in the depression. The awaiting male covers the eggs with a cloud of sperm.

Cartilaginous Fish

Sharks, rays, and skates are predatory fish with movable jaws, paired fins, cartilage skeleton, and unique a placoid-scale skin covering.

Predatory fish are built for speed; their bilateral body symmetry with paired fins and smooth skin is aerodynamically designed for quick bursts of speed in water. Combined with specialized nerve cells that connect directly to the olfactory bulbs in the brain, they are efficient predators. They have several evolutionary advancements:

- Gills conduct gaseous exchange; however, the adaptation is the ability to use two openings on top of their head to bring in fresh water to flush the gills when lying in a resting position—mouth down on the bottom. They do not need to swim to breathe.
- Like more complex animals, cartilaginous fish convert toxic ammonia to urea. Curiously, they retain high urea concentrations in their blood to simulate the general solute concentration in marine environments, so they never have to drink!
- Lipid storage in the liver adds buoyancy, which aids them from sinking and requires less energy to stay at a constant depth.
- Internal fertilization allows for greater reproductive success, although no parental care after birth.
- A lateral line of pressure-sensitive cells in a fluid-filled case that extends anterior to posterior and detects minute vibrations in the water.

Sharks are typical cartilaginous fish that are familiar to everyone who has seen the movie *Jaws*. Although typically not man-eaters, they have three specialized sensors that assist in the capture of their normal prey (fish):

- Their detailed sense of smell is particularly sensitive to blood and body fluids in the water.
- They are able to detect the faint electrical charges generated by the muscle movement of fish in their territory.
- Their lateral line system alerts them to any wave disturbance in the water, such as the submerged wake of a fish swimming nearby.

Bony Fish

Modern bony fish include catfish, trout, bass, tropical fish pets, swordfish, and tuna. Fossil evidence records their first existence around 40 million years ago. Bony fish are the most numerous, most recognizable, and most diverse type of fish. They have several unique adaptations that increase their complexity, including the following key evolutionary characteristics:

- The bony endoskeleton is reinforced by calcium carbonate for increased strength and weight.
- A *swim bladder* is a gas-filled sac that increases the buoyancy of the fish. It is thought to be an evolutionary modification of early fish lungs, because adding air to the sac increases buoyancy, and deflating it allows the fish to sink. This is a contrast to the light-oil lipid stored in shark liver for buoyancy.
- A hard protective gill plate operculum covers the gill housing on each side of the head. By repeated opening and closing of the operculum, fish can push water across their gills, allowing them to breathe without having to swim!
- The lateral line system is more complete than in the shark. They detect vibrations the same way humans hear music—by receiving and understanding vibrations in the medium.
- Scales are needed for protection and streamlining for speed.
- A four-chambered heart pumps oxygen-rich blood from the gills to all parts of the body.
- Kidneys and gills regulate water and salt balance.

- A brain with three areas, as in humans: cerebrum, medulla oblongata, and cerebellum (with an acute sense of vision and smell).
- A "complete" unidirectional digestive system: mouth, pharynx, esophagus, stomach, intestine, anus, which also includes liver, gallbladder, pancreas.

Amphibians

Amphibians are both terrestrial and aquatic, returning to water to reproduce. They were also the first land vertebrate. Interestingly, lobe-finned fish such as the lungfish are thought to be the precursors of amphibians; a mitochondrial DNA analysis completed in 1990 supports this hypothesis. In addition, embryonic analysis establishes that the forelimbs of amphibians are homologous to the pectoral fins of fish. Further, it is believed that the early amphibians breathed through lungs, also a fish adaptation. It is thought that competition for food and space may have compelled the early amphibians to the water's edge, forcing the colonization of new territories.

Amphibians have several key environmental characteristics that emphasize the transition from water to land:

- Metamorphosis from a mostly aquatic juvenile stage to mostly terrestrial adult form.
- Cutaneous respiration through moist membranes. This means their skin supplements the oxygen intake provided by gills and lungs and also limits their size. Most juveniles employ gills, whereas most adults have lungs.
- Double-loop circulation sends oxygen-poor blood from the heart to the lungs for oxygenation and then returns the oxygen-rich blood to the chambered heart for pumping to the rest of the body. Amphibians pump a mixture of oxygenated and deoxygenated blood, and amphibians are the first with two separate loops.
- Eggs are enclosed by n membranes, no shells, and will dry out if not placed in a moist area.

In moving from water to land, early amphibians probably feasted on the insects that were already well established there. Amphibians also adapted food-catching devices, such as a sticky tongue that is launched from the mouth at blinding speeds. Frogs and toads also evolved a skeleton and leg patterns that promote jumping, either for food or protection.

Frogs are probably the most common amphibians. Their metamorphosis from the larval tadpole stage, or hatchling, to the mature adult is interesting. The illustration *Frog metamorphosis* is a useful reference.

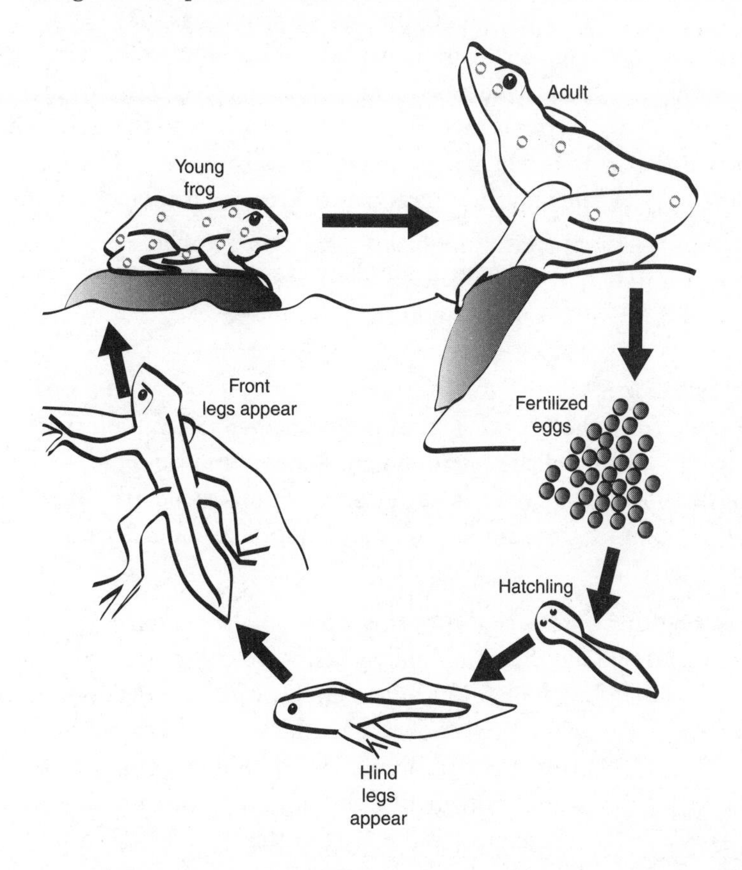

Frog metamorphosis.

Frogs always reproduce in wet areas, preferably water. Neither their skin nor eggs are water-tight, and both will dry out if exposed to air for prolonged periods. In temperate areas, when frogs end their hibernation, the spring air is filled with the species-specific mating call of the adult male. The familiar croak is generated when air is forced back and forth between its mouth and lungs, which vibrates the vocal cords both ways and is amplified by a vocal sac. Females respond only to the call of same-species suitors by entering their territory. The male mounts and embraces the female, often waiting hours or days for her to lay her eggs somewhere in the water in a gelatinous blob. The male then releases sperm directly onto the eggs, and the couple separates to live individual lives away from the eggs.

The fertilized eggs hatch into free-swimming larval tadpoles that look more like fish than frogs, with such fishlike features as a lateral line and tail fins. Newly hatched tadpoles feed on the stored energy in the yolk until their mouth opens, allowing them to eat algae. As they mature, they grow three sets of gills to supplement their respiratory surface, and they sprout hind legs, then forelegs. Next their tail is absorbed, and they exchange their gills for lungs as they begin to gulp air from the surface. The adult frog is carnivorous, whereas the tadpole is herbivorous; this dramatic metamorphosis is controlled by the hormone thyroxin and environmental conditions. The speed of metamorphosis is controlled by a combination of species-specific genetics and environmental conditions such as temperature and food availability.

Bionote

Certain amphibians such as salamanders do not proceed beyond the larval stage in their metamorphosis.

Reptiles

Based on fossils collected from the Carboniferous period (some 350 million years ago), an evolutionary link was established between amphibians and reptiles. Further studies in embryology and comparative anatomy confirm similar structures. It is believed that the earliest reptiles were similar to adult amphibians: four legs, small, with similar body symmetry.

Reptiles flourished during this time period probably because of the pioneer effect. They were more successful on land than amphibians and were entering a territory with plentiful food sources, with no predators, and where their only competition came with them. By the Mesozoic era, about 250 million years ago, reptiles were so diversified and dominant that the time span is often called the "age of the reptiles." During this time, the "terrible lizards" or dinosaurs roamed the earth and grew to amazing sizes, some evolved structures that enabled them to fly, and others recolonized the water.

Bionote

The pterosaur dinosaurs evolved the capacity for flight (pterodactyls, for instance). This species became extinct, and no modern-day reptiles can fly.

Reptiles have several key evolutionary characteristics that separate them from amphibians:

- Internal fertilization, which allows the female's reproductive system to develop a cover for the embryos with protective membranes and a shell
- The ability to conserve water by converting nitrogen-rich water into uric acid; this prevents evaporation due to the watertight skin layer
- Amniotic eggs

- Strong skeletons with legs located under the body to support movement on land
- An ectothermic metabolism
- Respiration via lungs
- Larger cerebrum than amphibians

The development of an amniotic egg was a major step forward in land colonization. All reptiles that lay eggs produce an egg with a watertight shell and its own food and water supply. Two membranes, the amnion and chorion, protect the embryo. The amnion, yolk sac, and allantois serve as a permeable membrane that allows gaseous exchange: oxygen in, carbon dioxide out. The developing embryo absorbs nutrients from the yolk via blood vessels extending from the embryo's gut into the yolk sac. Finally, the waste products of maturation are separated from the embryo and stored in a membrane-lined space called an allantois. In a sense, all four structures have combined to create an environment that provides for every requirement of the developing embryo. This self-sufficiency enabled reptiles to reproduce and inhabit the land.

The resting metabolism of a reptile is too limited to provide the internal warmth necessary to support its life function, so reptiles must move to areas where they can absorb, or lose, heat in relation to their surroundings. At one time, taxonomists classified them as "cold-blooded" because they cannot adjust their body temperature. A better term is *ectoderm*, because they are able to warm their blood by absorbing heat from the sun and other warm places.

Ectotherms have an advantage because they do not convert some of the energy gained from their food to heat, as humans do, so they are not required to eat as much. However, on the downside, the proper body temperature is important because cellular reactions do not occur in cold temperatures. So reptiles alternate between a sunny and a shaded area to control their body temperature. As the weather becomes colder, they are not able to maintain their normal metabolic rate, so they appear sluggish and lethargic. Their inability to function in cold temperatures forms the boundary for their existence. In fringe territories, reptiles are forced to hibernate in the winter.

Crocodiles exhibit an interesting reproductive phenomenon that characterizes reptiles. There are at least 20 species of crocodile that survived extinction and inhabit tropical and subtropical zones, including the southernmost parts of the United States. They are the closest-living relative to the dinosaurs. Interestingly, they provide care for their offspring for up to two years, which is uncharacteristic of most reptiles, who provide no support.

After fertilization, the female crocodile constructs a nest of twigs, branches, and organic plant matter. She then deposits the fertilized eggs in the middle and covers them with more nesting materials. The warm sun, tropical climate, and decaying vegetation provide heat for incubation of the eggs. The eggs are vulnerable to omnipresent predators, so the mother remains on guard in the vicinity. When the eggs have hatched, she transports them to their aquatic home to begin their growth phase.

Curiously, research on crocodile, alligator, and certain turtle eggs demonstrates a connection between ambient temperatures and the sex of the offspring. In most cases, hot seasons favor the birth of males. As in the case of crocodiles and most reptiles, the female creates a protective shell around each egg; however, they are not always hard like chicken egg shells, leaving them more vulnerable for predation. This type of egg production is called *oviparity* and is the simplest type.

To minimize the risk of predation, females of certain species keep the eggs internal until they hatch or just before they hatch in a process called *ovoviviparity*. The most human type of reptilian reproduction is called *viviparity*, because no shell is formed and the female provides nourishment for the young through a *placenta*. The placenta is similar to humans in that blood vessels from the mother transfer food, water, and oxygen across the placental membrane in exchange for metabolic wastes.

Reptiles colonized land beyond the capabilities of less-complex animals. The four-chambered heart in alligators, the complete lung system, and amniotic eggs are evolutionary breakthroughs carried forward into more complex phyla.

Birds

Fossils of birds called *archaeopteryx* were located in Germany and date to 150 million years ago. Archaeopteryx resembled a small dinosaur with wings, predatory teeth, and a bony tail. In fact, some taxonomists refer to birds as "warm-blooded dinosaurs with feathers." Modern birds are endothermic, feathered animals with two modified limbs for walking and two modified for flying. Although it may have resembled a flying reptile, Archaeoppteryx had feathers, one of the characteristics of birds:

- Feathers
- Flight
- Lightweight skeletons
- Efficient respiratory and circulatory systems
- Forelimbs modified into wings
- Endothermic metabolism
- Oviparity, incubated amniotic eggs
- Beak, no teeth

Feathers replaced scales as the body covering in birds and serve two primary functions: They insulate the body and provide lift for flight. *Powder feathers* produce a hydrophobic powder that repels water; when used in combination with the waterproof oil produced in specialized preen glands, water birds are able to keep their down and *contour feathers* dry. *Down feathers* are typically found in juvenile birds because the feathers are highly branched and soft, providing comfort and insulation. Down feathers grow underneath and sometimes between contour feathers in adults. Contour feathers represent the adult's plumage. They are large and cover the body, wings, and tail. Strong contour feathers on the wings and tail provide the lifting force and balance needed for flight.

Two hypotheses attempt to explain the evolution of flight. The first hypothesis states that winglike structures were used by reptiles to provide balance while capturing prey and may have also been involved as a weapon in the submission of prey. It is suggested that over many generations, the wings became large enough to trap smaller animals and eventually large enough to support flight. Another hypothesis begins the evolution of flight in trees, as tree-dwelling reptiles learned to jump between trees in pursuit of food or to escape predation. Winglike structures that allowed them to glide rather than fall may have provided a reproductive advantage. The act of flapping wings to increase thrust probably began shortly thereafter.

The act of flying for birds is similar to airplanes in that both require lift and thrust to overcome drag and gravity. Lift is created when the air pressure beneath the wing is greater than the air pressure above it, for both birds and airplanes. In both cases, the wing is designed to act as an airfoil, like the sail on a sailboat, to create high- and low-pressure areas. Bird wings are flexible and can be arched to create a greater airfoil, and they can orient the front of the wing upward to increase the angle of attack and resulting pressure gradients. Airfoils, such as bird wings, only work when air is passing over the wing, so either the bird has to be moving or wind must be blowing over the wing. This is the same reason that pilots prefer to take off by flying into the wind.

In the absence of wind, birds flap their wings to create thrust. The wings are fully spread during the down stroke to push air down and behind the bird. This creates a high pressure under and behind the wings that lifts and propels the bird. The wings are oriented on the upstroke to block as little wind as possible. Aside from feathers, birds have several modifications that enable them to fly:

- Hollow bones with air sacs
- Keeled sternum, or breastbone, that attaches strong breast or flight muscles
- High metabolic rate
- Efficient respiratory system

Most of the flight features are also key evolutionary characteristics, because birds are built for flight. For instance, most bird bones are thin, hollow, and fused to make a rigid lightweight skeleton, including an oversized sternum. Likewise, birds and mammals need a high metabolic rate to generate body warmth. As a result, endotherms, such as both birds and mammals (humans), have to eat frequently to turn food energy into heat energy. In general, birds maintain a higher resting body temperature, 104°F (40°C), than humans do, 98°F (37°C).

Flight and metabolic requirements stress the circulatory and respiratory systems, and as a result, birds developed improvements in both areas. The avian respiratory system exhibits a large surface area like the reptiles', combined with the countercurrent exchange and unidirectional flow characteristic of fish, to create an efficient air-processing system. Inhaled air passes into the two bronchiole tubes, where some of the air is funneled directly into the lungs; most of the oxygen-rich air is shunted to nine posterior air sacs that serve as containers. Exhaled carbon dioxide–rich air leaves the lungs continuously and is shunted into anterior air sacs, while the oxygen-rich air in the posterior air sacs continually flows into the lungs. Anterior carbon dioxide–rich air is exhaled into the atmosphere. The net effect is a constant flow of oxygen-rich air flowing over the lungs, which greatly increases the absorption rate by the lungs. In addition, the circulatory system is pushing the oxygenated blood in the opposite direction, thereby continually creating a low oxygen concentration in the blood that diffuses oxygen from the lungs into the bloodstream.

Bionote

Normal human hearts beat about 60 to 80 times every minute, compared with 600 beats per minute for a hummingbird.

Like crocodiles and humans, birds have a heart with separate ventricles that keep oxygen-rich blood headed for the body separate from oxygen-poor blood headed for the lungs. This greatly increases the oxygen supply to the cellular mitochondria, so as to not limit cellular respiration. Further, the rapid heartbeat of most birds pushes the blood to support their unique metabolic rate.

The digestive and excretory systems are designed for the efficient and quick digestion of food and the elimination of wastes to support their high metabolic rate. Because birds don't have teeth, seeds and other tough foods are cracked and broken in the two-part stomach. In the *proventricles* chamber, stomach acid and enzymes combine to begin the chemical digestion of the food that then moves into the *gizzard.* The gizzard is a muscular chamber that usually contains sand and constantly churns the stomach to mix and churn the food. The contents then move quickly to the small intestine, where bile and pancreatic enzymes further digest the mixture. The nutrients are continually absorbed in the intestinal bloodstream. Undigested food is eliminated along with uric acid from the kidneys' detoxification of nitrogen, via the cloaca, a common excretory-elimination exit. Bird droppings are a combination of uric acid and fecal material.

Reproduction in birds is a seasonal activity controlled by a combination of programmed DNA events and prevailing environmental conditions. Like humans, male birds have two testes that produce sperm that travels through ductwork into the cloaca. Females release eggs from the single ovary into the *oviduct*, where they are fertilized. As the fertilized egg travels down the oviduct, membranes and a shell are added before entering the cloaca that transfers the egg to the nest. The oviparity feature of the parental care of the amniotic egg is demonstrated by the lengthy time the birds brood their offspring. See the illustration *A typical bird.*

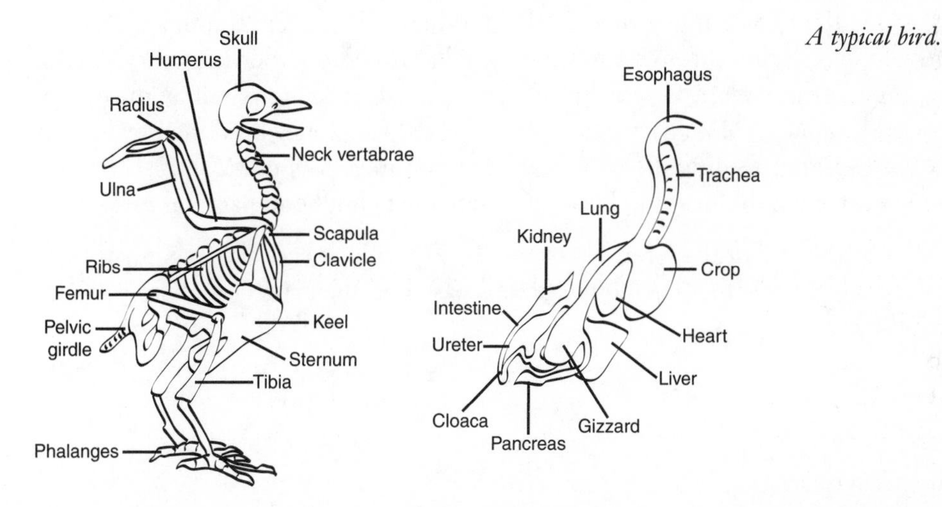

A typical bird.

Birds exhibit many reptilian features, such as amniotic eggs and body symmetry, but they are considerably advanced in most organ systems. Fossil evidence establishes a link between reptiles and birds and also documents a massive bird extinction that occurred when the dinosaurs became extinct about 65 million years ago. Today's birds are the descendants of the survivors of the genetic bottleneck event.

Mammals

Fossil records document a chain of reptiles with ever-increasing mammalian characteristics that eventually become more mammal than reptile about 225 million years ago, before the age of dinosaurs. The fossil records also indicate that the first mammals were probably small, like a mouse; endothermic; and, judging by the size of their eye socket, nocturnal. Large eyes allowed better night vision, and the endothermic metabolism allowed them to function during the cool of the night—when reptilian, ectothermic dinosaurs were inactive—thus minimizing competition and predation.

Following the extinction of the dinosaurs, the mammalian population exploded, and their genetic variation increased dramatically, showing a clear separation from reptilian construction. Although there are numerous common mammalian features, the two key evolutionary characteristics are mammary glands and body hair. Both are and described in the sections that follow.

Mammary Glands

Almost all mammals are viviparous, meaning that the offspring develop within the mother and are then born fully formed without a shell. The female continues the nurturing of the offspring through a nutritious milk, produced in her *mammary glands*, that contains all of the water, fats, protein, minerals, and sugars required by the offspring for early growth. Mammary glands are required for all newborn mammals until the mother stops feeding them or *weans* them on to solid food. Mammals are also unique in the vertebrate world because of extended childhood parental care.

Sometime near the end of the Cretaceous period, 75 million years ago, mammals had evolved to form their distinct taxonomic groups based on their method of embryological development:

- Monotremata
- Marsupial
- Eutherian (placental)

The monotremes are the simplest mammal and are represented by only three species found in Australia or New Guinea. This geographic reproductive isolation likely contributed to their rareness. Monotremes also exhibit several reptilian characteristics, such as laying eggs with shells (oviparous) and a similar pelvic skeletal structure, and like birds, have a cloacalike single opening for waste and reproductive substances to exit the body. However, they do share the mammalian characteristics of mammary glands and hair. Curiously, the young lap milk from the mother's hair rather than nursing because the mother has no nipples. The duck-billed platypus is the most common example and is a genuine oddity. They have a broad, flat, almost hairless tail that matches a protruding ducklike bill, and they have webbed front feet to support their semi-aquatic lifestyle. Imagine, an egg-laying, duck-mouthed mammal!

Marsupials are unique as well. Their young develop as viviparous offspring internally for a brief period before their birth, often within a week of fertilization. The unfinished newborn must locate the mother's abdominal pouch without any parental help and seek out her nipple to receive the life-saving milk. They remain in the pouch, usually attached to a nipple, until they are completely formed and ready for independent living. Like monotremes, most marsupials such as kangaroos and koalas are native to Australia and New Guinea, a reminder of the breakup of Pangaea 70 million years ago. The notable exception is the opossum, which is found throughout temperate areas such as North America.

Bionote

The introduction of dogs and other placental mammals into marsupial territory has been disastrous for the marsupials because of increased competition and predation.

Eutherians are more commonly called placental mammals because their placenta is unique and separates them from the other mammals. The placenta is a membranous organ that permits diffusion of all nutrients, water, and oxygen from the mother's blood to the developing embryo, with metabolic wastes returning through the placenta. The placenta allows for a prolonged, protected, and personal gestation period longer than other mammals. The viviparous embryology prolongs the development to a point where newborns such as horses and deer can stand up and even run only a few hours after their birth.

Body Hair

Mammalian hair serves many functions. Most agree that the primary function is for insulation to protect against temperature changes and to minimize the metabolic energy that has to be extended for internal heat production. The *underhair* for a fur-bearing mammal is short, thinner, and more densely packed to provide maximum thermal protection. *Guard hairs* are the longer, coarser hairs that provide mammals with their coat color. Coat color is significant for a number of reasons such as camouflage and mating preference. Modified hair is used for whiskers that enable mammals to increase their sensory input in low-light areas, at night or when they're underground. Curiously, another type of modified hair is the stiff and sharp shafts of the porcupine and hedgehog

Bionote

Bats are the only mammals capable of unassisted flight.

quills. Their defensive posture and capabilities are well known and, contrary to some belief, they cannot launch or throw their quills!

Other Unique Mammalian Qualities

Mammals have unique teeth. Most vertebrates have a constant supply of new teeth, but mammals receive only two sets: a first set, called baby teeth, which is replaced in juvenile stages by permanent teeth. It has been said that one can determine what a bird does for a living by looking at its feet and beak; likewise, in a mammal, look at the teeth. Adult teeth feature prominent canines that are used for piercing and indicate predator; these are usually accompanied by sharp, shorter incisor teeth in the front for biting. Along the jaw, flatter and thicker teeth, called molars and premolars, in plentiful supply indicate herbivores. Humans have a balance of all four types, indicating omnivore; we eat almost anything.

Most other organ systems established in less-complex animals are basically the same in mammals except body style. Structural modifications allow loose skin to drape over wings for flight. Whereas whales have forelimbs modified into fins and no hind legs, their slick, nearly hairless body is streamlined for swimming. Elephants are not as big as the largest whales, but they have adopted heavy bone structures to support their massive weight on land. They have huge ears for long-distance sound sensitivity and have two elongated incisors modified as tusks. Rabbits have longer and stronger back legs than front for jumping, whereas cheetahs have a supple spine for speed bursts. Mammals are adapted for almost every venue.

The most common mammal is the *Homo sapiens.*

Homo Sapiens

Three key evolutionary characteristics mark the arrival of *Homo sapiens:*

- Skeletal alignment to support upright structure and two-legged walking
- An enlarged brain that prompted the development of tools and shelters
- Community organization that led to prolonged time for childcare, which began culture

Much of what we know about early, ancestral prehumans comes from fossils in Africa and Israel. In 1924, a skull was found in South Africa that was different from previous worldwide discoveries and was dated to be approximately four million years old. Because of its uniqueness, it was classified in a new genus, *Australopithecus,* meaning "southern ape."

The combination of fossils discovered, including pelvic structures, during this time led researchers to believe that these animals walked upright on two legs. They gave the species the name *hominids.*

In 1974, an expedition led by Johnson and White discovered a more detailed skeleton of and adult from the same time period that they named "Lucy."

In 1977, Dr. Mary Leakey discovered fossilized hominid footprints, probably an adult and a juvenile, from the same time period, further confirming that the hominids walked upright on two feet. Additional remains have been located, and the various pieces, like several giant jigsaw puzzles, have been pieced together to create an image of an organism that is short in stature, nearly 3 feet tall (about 1 meter), and walks upright. Hominids are not classified as humans, but they do have many similar identifiable features.

Dr. Richard Leakey discovered yet another type of skull; it had a much larger brain case and a smaller face. Implements made of fashioned stone and carved bone were also found with the skeletons, indicating a higher level of development. In fact, based on this evidence, it was decided to place the new species in the same genus as modern man. He became *Homo habilis*, or "handy man," referencing the tools found in conjunction with the remains. At this same location in Kenya, what appears to be a settlement or community-gathering site was located along with more tools and animal remains. It is not known why *Homo habilis* became extinct.

Bionote

Upright walking appears to predate a large brain. According to Dr. Stephen Gould, "Mankind stood up first and got smart later."

Homo erectus, "upright man," an even larger-brained species, appeared in fossil finds throughout Africa and parts of Asia and Europe.

To survive the climactic change, *Homo erectus* had to create protective shelters, thus providing an increased sense of community. From archaeological digs in France estimated at 500,000 years old, primitive shelters were discovered that contained more advanced tools, indicating that *Homo erectus* also mastered fire and used it for his own purposes, such as warmth and alternative food preparation.

Later discoveries in the Neander Valley in Germany indicate that a new hominid with even greater mental capacity appeared almost 150,000 years ago: *Homo sapiens neanderthalensis*, nicknamed Neanderthal for the discovery location. They were the first to structurally appear similar to modern man. They were short, heavily muscled, toolmakers, and engaged in rituals, such as burials, that indicate an understanding of abstract concepts.

Approximately 100,000 years ago, another species appeared on Earth with characteristics and appearances even more closely resembling modern humans. Cro-Magnon fossils indicate that they originated in Africa and colonized the rest of the world. They were known to overlap territories with the Neanderthals and may have interbred. However, the Neanderthal lineage died out, and the Cro-Magnon continues. It is not known whether the Neanderthals were absorbed by the Cro-Magnon or defeated in battle or competition by the mentally superior species. Cave artistry displays a community culture with sophisticated tools, sharpened cutting points probably for weapons, and the time and talent to create artwork such as cave drawings. Accepting this evidence would lead one to think that we are descendants of the Cro-Magnon phylogeny.

Bionote

Mitochondrial DNA analyses indicate that current inhabitants of central Africa have greater genetic variability than people anywhere else on Earth, supporting the hypothesis that life originated there.

The communal aspect of living at this time probably enhanced the division of labor that remained stereotypical for a long time. The male was usually bigger and stronger and became the food slayer and household defender. They also knew the advantage of cooperating. Even during *Homo erectus* time, evidence exists of group cooperation at a massive animal kill. The females were the birth makers and had the time and protection to provide extended child care for the infants. During this time, stories, knowledge, and advice were passed from parents to offspring for generations, creating the culture and establishing behavior for the society.

The modifications that created a group of animals with a backbone also include both very simple and very complex organisms. The simplest vertebrates are structurally very different from the more sophisticated specimens. However, at some point in their development, they have many common characteristics.

The Least You Need to Know

- Fish are the largest and simplest group of vertebrates and contain members that are considered to be the most primitive vertebrates.
- Amphibians share many homologous structures with lobe-finned fish.
- Most amphibians have an herbivorous, aquatic juvenile stage and a terrestrial, carnivorous adult stage.
- Egg-laying strategies vary according to species.
- Bird structures are highly adapted for flight.
- Mammals have body hair and mammary glands.

Part 6

Biosphere and Ecology

The concepts, theories, and ideas presented in the previous sections of the book are synthesized in this part to create the overall picture of how life and nonlife interact. If you have studied the previous chapters and completed all the Bioproblems, you are ready. Actually, this part is less-hardcore science than the other parts because it focuses on the big picture rather than the molecular level.

Chapter 23

Biomes

In This Chapter

- What affects a biome's overall health
- Plant and animal adaptations in response to environmental resources
- Man's influence over the biomes
- Terrestrial and aquatic ecosystems
- How climate affects a biome

As we examine individual climatic ecosystems, you'll find that their stories are revealed in the plant and animal life that live there as well as how they interact with their environment. Each ecosystem has specific characteristics that weigh upon all creatures that inhabit a territory.

Specific plant and animal modifications are presented in a manner that addresses the configurations of the environment. For example, you'll better understand why the fur of the arctic hare turns white in the winter as well as that of the arctic fox, its predator. This chapter also explains why you wear a coat in the winter … but it doesn't explain why you choose to live in an area that is too cold or too hot.

Biomes and Terrestrial Ecosystems

The earth is a unique and fascinating place. Astronomical events such as rotation and revolution create circadian cycles and influence weather. The tilt of the earth in respect to the sun creates seasons, large bodies of water affect climate, and the atmosphere creates and moves weather. These abiotic factors unevenly applied on Earth create differing ecosystems called *biomes.*

A biome is a geographic region characterized by distinctive climates and specific kinds of plants and animals. They may exist in more than one location around the earth, but similar biomes have similar organisms with similar adaptations as natural selection favors those who match the environment. The interaction between plants and animals in a particular biome for long periods of time often leads to coevolution. Refer to the illustration *Biomes.*

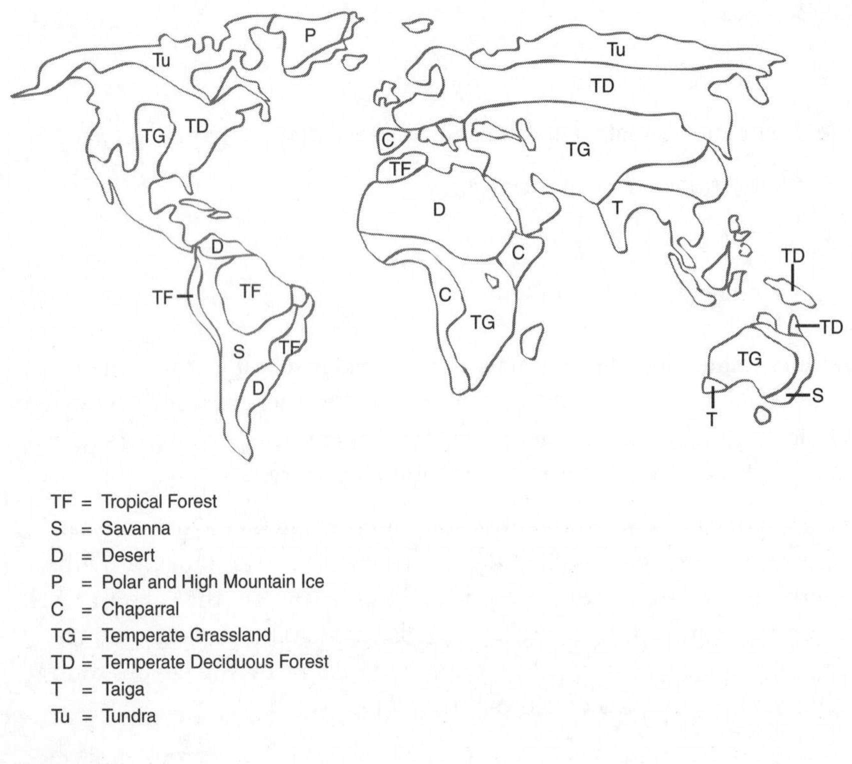

Biomes

Each biome is a large terrestrial *ecosystem* that usually contains differing yet closely related individual ecosystems. Biomes are classified primarily based on the types of vegetation that grow there, which in turn depends on the climate, especially the amount and timing of rainfall. Refer to the illustration *Biomes and climate.*

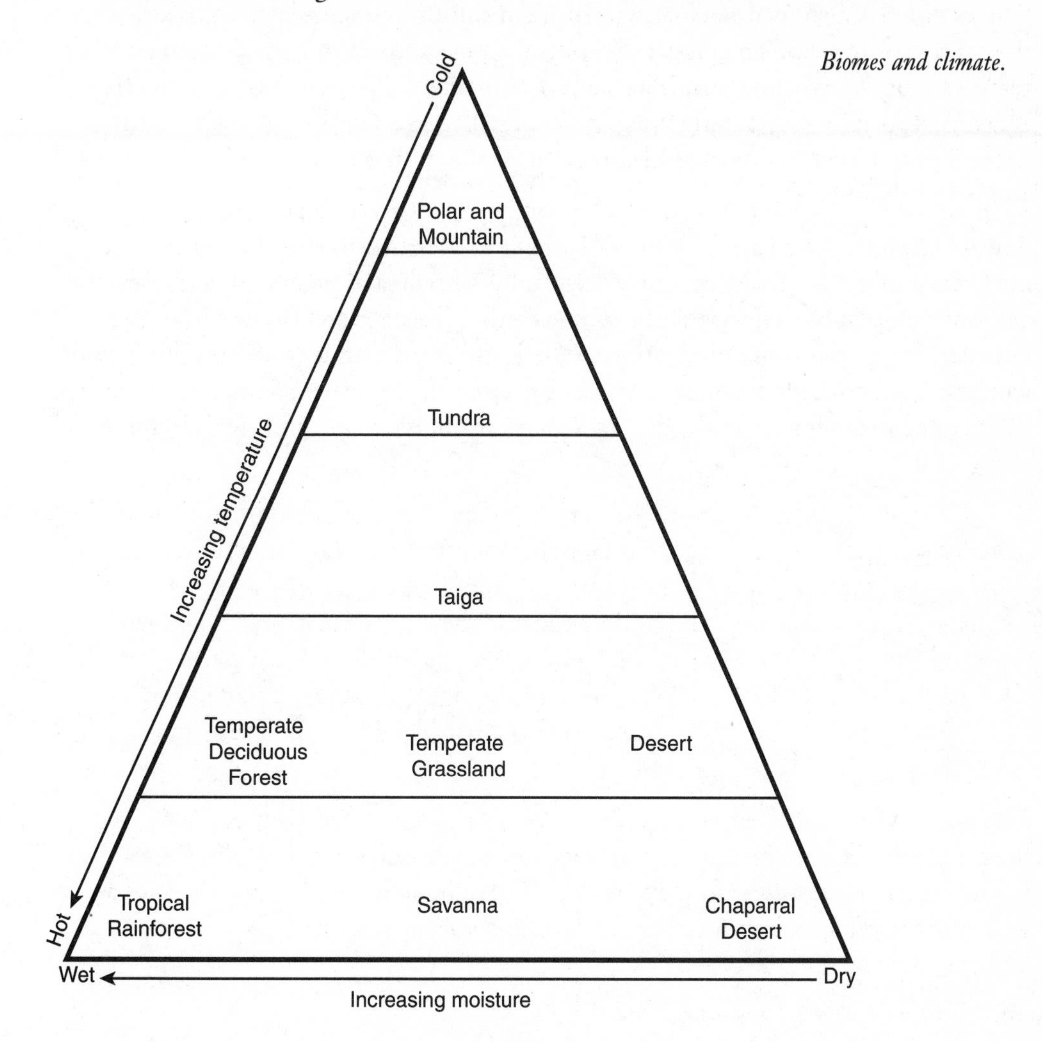

Biomes and climate.

An ecosystem includes all of the abiotic factors, such as temperature, sunlight, and other nonliving factors; and biotic factors, including all of the living elements, such as the types of trees and the number and type of birds in an area. The earth is divided into nine separate land biomes or terrestrial ecosystems: chaparral, desert, polar and mountain, savanna, taiga, temperate deciduous forest, temperate grassland, tropical rainforest, and tundra.

Chaparral

Chaparral is Spanish for "place of always green shrubs," which is also a good description of this ecosystem. The climate is controlled by cold ocean currents that produce temperate but rainy winters with hot and dry summers. Because of the ocean effect, chaparral ecosystems are found along cool, costal areas such as coastal Chile, the Mediterranean, and mid-coast California, normally in the latitude band of 30° above and below the equator.

The dry season spawns periodic fires that tend to destroy invasive species but normally do not kill the hardened scrub oaks and other adapted plants. In fact, the meristematic areas of several shrubs are somewhat fire resistant and benefit from the nutrients released by the action of fire. Others are fire dependent because their seeds will only germinate after exposure to a hot fire; some even have flammable oil in their leaves to promote burning. The leaves are also leathery, tough, and small to minimize water loss.

Bionote

The leaf-oil gives the sage leaf and bay leaf, two typical chaparral plants, their unique culinary flavor.

The climax vegetation for this biome limits the animal population to brush browsers such as deer, seedeaters such as mice, and fruit eaters such as birds. Predatory hawks and snakes are abundant, feeding on the small herbivores. Chaparral animals specialize in camouflage; most are muted shades of brown or gray that blend into the background, so when they're standing still, they are often hard to see.

The chaparral climate is especially pleasant because of the slight temperature variation. The impact of human migration into these areas such as the Mediterranean perimeter and parts of California may spell loss of habitat for the native species as well as other changes commensurate with human occupation.

Desert

Deserts are regions that receive less than 10 inches (25 centimeters) of precipitation per year. In equatorial areas, they may also be hot; however, high-plateau deserts are common and have moderate, even cold temperatures. The largest desert in the world is the Sahara desert in North Africa, which averages less than about 1 inch (2 centimeters) per year of rainfall. Some years the Sahara receives no rain, which helps to explain why it appears mostly as a lifeless, barren landscape.

Certain plants and animals have adapted to the dry conditions of the desert. Water collection and retention are as difficult as when aquatic plants first became terrestrial.

CAM are specialized plants that regulate their stomata, cacti have modified leaves, and succulent plants have a waxy cuticle to minimize evaporation. In desert areas, such as the southwestern United States, barely enough precipitation falls to allow these plants to survive. Most desert plants absorb the small amount of rain, grow, and reproduce very rapidly to capitalize on the short growing season. They are also prolific seed producers. The seeds are an important species-sustaining feature because they are able to delay growth until the growing season begins.

Animals have adapted to the dry conditions through structure and behavioral modification. Most are nocturnal or avoid the drying sun and wind whenever possible by burrowing into the earth or inhabiting caves or crevices. Spade-foot toads escape the heat by a process similar to hibernation called *estivation.* They burrow into the ground until the hot, dry season has passed and then emerge to eat and reproduce. Several species such as the desert kangaroo rat excrete highly concentrated urine to conserve water.

A simplified food chain almost always begins with the seeds, which are readily consumed by rodents. During the growing season, seasonal flowers are also consumed along with berries. Both the flowers and berries contribute their water content to the herbivore. Predatory animals such as snakes and hawks feast on the small herbivores. They, too, receive water from their prey. Water is not wasted by desert life.

Deserts are not static. As climates change and humans alter the environment, deserts can become fertile or they may spread, creating additional low-vegetative areas. In certain irrigated areas of the Middle East, Mexico, and southwestern United States, the desert soils and sunny days have combined to create a productive agricultural area. In vast regions of central Africa, however, overpopulation, overgrazing, and poor farming techniques have allowed the deserts to increase in a process called *desertification.* The resulting hardship on both plant and animal life creates extreme environmental conditions.

Polar and Mountain

Cold temperatures characterize polar and mountain ecosystems where the terrain is permanently snow or ice covered and is inhospitable to most life forms. The polar proximity or high altitude typical of these areas prevents heat penetration, so even during the summer, the temperatures seldom climb above the freezing point of water. Without a usable source of fresh water, the environment is often more severe than the deserts. The complete lack of vegetation precludes herbivore habitation, which normally forms the foundation for food chains. Any life in the area normally subsists in the ocean waters. Atypically, the South Pole does support populations of penguins, who mainly subsist on food harvested from the marine ecosystem.

Savanna

Savanna comes from the Spanish word for "meadow." It represents the transition area between the desert and the grasslands and is characterized as grassland with sparse tree growth—a meadow. Tree growth is sparse because of limited rainfall and frequent fires. The African savanna is the most well known, but large savanna regions also exist in South America and Australia. There is even a narrow strip running vertically north from east Texas into Lower Canada.

Plants and animals have adapted to the savanna's short, rainy season followed by the long, dry season. The growing season is long enough to supply the basic requirements to support herds of large herbivores such as zebras, gazelles, giraffes, bison, and deer, depending on the location. The climatic conditions allow the savanna to host the largest assemblage of grazing animals on Earth. Likewise, their predators are found in sustainable populations of large cats, wolves, and scavengers such as hyenas and coyotes.

Savanna plants adapted shallow fibrous root systems typical of most desert plants. The root system responds with growth bursts after a fire, thereby eliminating competition from invader species.

Curiously, the grazing herbivores avoid the dryness of the season whenever possible by migrating almost continually to areas where the grass has received enough rain to prompt the growth of tender new shoots. The rainy season is also the reproductive season for most savanna animals, as their biological clock marks the time for birth to coincide with lush, springtime vegetation.

Taiga

Taiga biomes are almost exclusively gymnosperm growth areas and are sometimes called coniferous forest ecosystems. The modified needles and protective cones of pines, hemlocks, spruce, and fir trees are environmental adaptations that suit their biome. Summers are short and mild. Like chaparral ecosystems, the dry summers invite forest fires, which actually serve to sustain the ecosystem by destroying deciduous trees, burning undergrowth, and increasing the amount of certain nutrients valued by plants. Additionally, fire-resistant cones open because of the intense heat and are the first to release seeds to reforest the area.

Typically the winters are long with a protective layer of snow that insulates tree roots but prohibits the formation of liquid water. Snowfall may be heavy, which adds to the overall amount of precipitation, but it may not be useful because of freezing temperatures in the winter and rapid snow melt in the spring. Curiously, the deep snow provides rodents, rabbits, and other burrowing animals cover from predators. Forest mice are able to tunnel through the snow to reach favored food areas without being noticed by predatory birds or cats. The snow pile also provides a layer of insulation and a windbreak. Rather than create hardships, the snow is often beneficial for some species.

The moose, wolves, hawks, and rabbits that inhabit the taiga and remain active are structurally adapted to the cold, with fat layers and heavy coats. Other animals have strategies such as the ability to hibernate, like bears, and miss the bad weather altogether. Hibernation resembles sleep but is a long-term state where the metabolic rate of an animal is greatly reduced, allowing the animals to pass through the extreme conditions and thereby avoid the stress. Hibernation is also unique because the hypothalamus uses hormonal control to lower body temperature and reduce the metabolic rate without harming the organism. The decreased metabolism allows stored fat reserves to last for the entire season. Other animals, like the Canadian goose, avoid the hostile weather by migrating to warmer temperatures.

Temperate Deciduous Forest

Temperate *deciduous* forests are found in the latitude band between 35° and 50°, where evenly distributed annual precipitation allows the growth of large trees such as oak, poplar, and maple. Most of the eastern United States, coastal Canada, and eastern Europe are located in this biome.

Temperature extremes create hardships for the life-forms in this biome. Although the climate permits abundant plant and animal species, summers are hot and winters are cold. Indigenous animals display some of the same techniques used by desert and taiga life-forms, such as hibernation, avoidance of harsh conditions, and seasonal fat storage. The hardwood trees have added another feature: They drop their leaves at the onset of shorter days and colder temperatures. This survival advantage prevents damage to tender tissue to minimize the need for water during the months it is likely to be frozen and unavailable.

Bioterms

A **deciduous** tree is a hardwood tree that drops its leaves in the cold season.

The climate is well suited for most plant growth, and over years, the forests create a telltale carpet of leaf litter that is a home or covering for many animal species. The leaf litter also protects the tree roots from freezing and decomposes to rich humus. The high humus content did not escape the attention of early farmers, who cleared virtually all the original forest for farmland and tree products.

Temperate Grassland

Temperate grasslands are found in the same latitude band as temperate deciduous forest, but the seasonal, often-severe droughts and predictable fires prohibit the growth of trees. Conversely, grasses can burn to the ground line and regenerate quickly because of meristematic growth areas in the root. This phenomenon continually reestablishes grasses as the climax species.

Grasslands exist in many areas. In Africa, they are called veldt; in Eurasia, steppe; in South America, pampas; and in North America, prairie. The combination of climate, soil type, and the protective, soil-enriching grass groundcover makes the soil quality in this biome the most fertile in the world and extremely productive for farming. The deep and extensive root systems of prairie grasses are credited with the buildup of humus. The roots overlap, creating an almost impenetrable mat that resists erosion and holds decomposing vegetation in place so the nutrients are not washed or blown away. In fact, most of the original grassland has been converted into heavy agricultural-use areas.

Grasslands support grazing animals such as sheep, horses, bison, pronghorns, zebras, and wildebeests that have adapted broad, flat molars for chewing coarse prairie grass. Interestingly, because of the looser soil types and seasonal frozen soil temperature, burrowing animals such as badgers and prairie dogs escape massive range fires and weather extremes in their earth shelters. They thrive, as do the predators, hawks, snakes, and coyotes.

Natural grasses that created the fertile soil are almost completely displaced by cash crops such as corn and beans that are harvested each fall, leaving the soil exposed to the elements for the remainder of the year. A reoccurrance of the Dust Bowl of the 1920s and 1930s looms on the horizon as more natural protection is removed.

Tropical Rainforest

When the weather is warm and humid all year with rainfall measured between 200 to 400 centimeters, *tropical rainforests* dominate the ecosystem. Because most are equatorial locations, the amount of actual daylight contributes to the lush vegetable growth; daylight seldom goes below 11 hours per day and may average much longer lengths. Because of the combined contributions of these abiotic factors, the plant growth is luxurious. Trees often grow to a height of 200 feet (60 meters) and completely close in the forest, so little direct light penetrates to the ground. The multilayered canopy created by the various layers of plant growth establishes vertically stratified territories for animals to occupy. One of the reasons that the tropical rainforest contains the greatest number of different species is because of the number and variety of potential habitats. The animals are not limited to just below-, on-, or above-ground locations. The immense forest canopy with layers of vegetation provides more arboreal habitats than any other area in the world.

Animal life is particularly abundant, varied, and often beautiful. Most animals avoid the forest floor, where the large predators lurk, instead choosing an arboreal or aerial existence. Colorful insects, luxurious-looking birds, and small mammals occupy the canopy area, whereas amphibians, reptiles, and larger mammals inhabit the ground. Adaptations in the jungle are often more cosmetic than integral. Certain tree-frog species are brightly colored to attract mates and warn predators that they taste incredibly bad.

CAUTION

Biohazard

The tropical rainforests contain the greatest biodiversity in the world. It is thought that potential cures for diseases and other human needs may be hidden in the massive number of lifeforms that are found only there. Unfortunately, the tropical forests are being cleared in most developing countries to create more farmland and living space!

Tundra

The cold tundra biome is located inside the Artic Circle between the polar mountain and taiga. The word *tundra* comes from the Russian word for "marshy plain," which is a favorable description for this ecosystem. The climate is not only cold, the polar latitudes also alternately create extremely long days and nights. During the long summer days, the top layer of soil thaws enough for rapid plant growth. However, beneath the thin ribbon of thawed topsoil, the impermeable *permafrost* layer prevents the liquid water from percolating down into the subsoil. Because of this, the water stands in pools, as in a marsh.

Plants have adapted to the tough conditions by producing shallow, fibrous roots that absorb water quickly and provide an anchor against the artic winds. Mosses, lichens, and some tough grasses sustain themselves in these harsh conditions and provide grazing for herbivores.

Birds and herds of herbivorous grazing animals migrate into the tundra during the growing season to enjoy the brief period of vigorous plant growth and the seasonal insect hatching. Predatory wolves and other herd followers prey upon the migrating herds to limit their population. Several year-round residents have special adaptations. Small mammals burrow into underground shelters. The arctic fox and arctic snowshoe hare change coat color to match the snow, and everything adds an extra fat layer for insulation.

The tundra is one of the most fragile ecosystems, and it is slow and laborious to reclaim disturbed areas of the tundra. Conversationalists fear that the discovery of oil in these areas may bring an end to this sensitive biome.

Aquatic Biomes

Water covers approximately two thirds of the earth's surface and is so large that it exerts a tremendous influence on local and worldwide climate patterns. Water evaporates from the ocean and falls on land. Wind created on the ocean-atmosphere interface or ocean-land interface pushes weather patterns across the earth. Most of the oxygen production is generated by ocean-borne algae along with the removal of carbon dioxide to continually cleanse the air. There are three types of aquatic ecosystems: estuaries, freshwater, and marine.

Estuary

An *estuary* is a location where saltwater mixes with freshwater to create variable salty, *brackish* water. Most estuaries are roughly one-half as salty as the oceans, which allows certain species that are predominantly freshwater or saltwater to visit or inhabit these unique ecosystems.

Estuaries are highly productive because they are normally shallow, allowing photosynthesis to occur with plants rooted on the bottom. They also receive regular tidal flow from the oceans or seas and current flow from freshwater streams or rivers. They rival tropical rainforests in their number of species. The aquatic life and bird populations are enormous. The biotic factors create ideal growing habitats for temperate aquatic plants, such as plankton, which multiply during the growing season. This begins a long food chain that exemplifies the prosperity of the region.

Both plants and animals have adapted to the slight salt-concentration variation that occurs whenever a torrential rain or massive snow melt upstream sends thousands of gallons of freshwater into the estuary, or conversely a drought temporarily increases the salt concentration. Hurricanes and tropical storms can also add to the salt concentration as they blow ocean water into the estuary. The variation in salt concentration forces adaptations by the plants and animals. The mangrove tree solves the salt problem with its specialized glands that secrete excess salt through its leaves.

Most of the world's largest cities, such as Tokyo, Baltimore, and New York City, are located on estuaries. The impact of urbanization on estuaries has always produced decreased productivity in fishing harvests until protective measures were installed and enforced. In earlier times, industrial and human wastes were often piped directly into the estuaries, creating a nutrient imbalance and chemical invasion. Sensitive algae and other plants diminished in numbers, which affected everything above them on the food web. A researcher at the Academy of Natural Science in Maryland described the Patuxent River estuary as an "underwater desert."

Conservation procedures appear to be helping in certain areas. The elimination of the bioproductivity (the ability to create living organisms) of the estuaries affects numerous food webs and neighboring systems. It is similar to the destruction of the tropical rainforests.

Freshwater

Water is considered "fresh" when the salt content is 0.005 percent. Freshwater normally begins its journey as either mountainous snowmelt or accumulated rainwater. For fast-moving, snowmelt mountain streams, the water is cold and clear because the current is too fast to allow a buildup of sediment. Inorganic nutrients such as phosphorus and organic nutrients such as dislodged vegetative material wash from inaccessible mountains or polar biomes into a more productive area by the action of these small, swift streams. The only aquatic plants are algae types that can cling to rocks or other debris. Invertebrates are sometimes able to gain a foothold in slow-water eddies and behind natural dams. The biodiversity is limited because of current speed and cold water temperatures. The plant and animal life near the water are often well developed and plentiful.

Farther downstream, trout orient themselves into the current to feed on invertebrates trapped in the flow. As the topography levels and small streams join to create larger, slower-moving rivers, they accumulate more sediment and debris for habitat and food source. This bounty also attracts more predatory fish such as bass and other predatory birds and mammals.

As rivers reach their destination, the topographic relief flattens to create a slow-moving river, which has an abundance of sediment. Accumulated natural and man-made runoff from upstream concentrates in the lower river to magnify available nutrients. Photosynthetic plankton, algae, and rooted plants flourish unless the water is poisoned by the runoff. The water is usually cloudy because of the suspended sediment load and free-floating plankton. In extremely murky water, certain fish species such as catfish employ their modified sensory tissues to locate food by scent or taste. Bottom-dwelling mollusks are prevalent unless a heavy sediment load buries them. Plant and animal life along the edge is plentiful.

Freshwater lakes and ponds are often nutrient rich because of surface runoff and the natural life cycle of living things that crowd together in shallow water. This nutrient-loaded area is called the *littoral zone* and is the productive area of nonmoving freshwater. In deeper water, sunlight provides photosynthetic opportunities to free-floating plants at a depth determined by the clarity of the water. In some cloudy lakes, the photosynthetic layer is measured in inches.

As open-water organisms complete their life cycle and deposit on the bottom, their components are recycled by a host of scavengers and bacteria. These decomposers utilize oxygen as they recycle the accumulation. In heavy decomposition zones, the recyclers often consume all the available oxygen. The *benthic zone* is often oxygen deficient and prohibits air-breathing animals such as fish from entering. This stratification is compounded in the warmer months as the sun creates a warm upper layer of water that is lighter and does not mix with the colder, denser, deeper water. This creates nutrient deficiencies in the upper layers that may decrease photosynthesis. Fortunately, during the colder months, the surface waters cool to a lower temperature than the bottom water, making them denser. So the now-denser surface water moves to the bottom, forcing the less-dense bottom water to the top. This "lake turnover" brings sediment and nutrients to the surface for use by aquatic plants. Warm mountain lakes can turn over when a rush of cold snowmelt enters the lake and pushes the warmer water off the bottom. Even clear mountain lakes, called *oligotrophic lakes* because they contain small amounts of organic matter, contain enough nutrients to sustain a photosynthetic population. The exception to this rule is the ultra-oligotrophic lakes that are too cold for most life-supporting activities.

Plants and animals have modified structures that enable them to survive in the environment in which they are located. For instance, algae cling to rocks and mosses using *rhizoids* as they and mayfly nymphs hold on against fast-moving waters. Trout are cold-water adapted, but farther down the stream, warm-water species overlap territories. Cattails, arrowhead plants, and seeds thrive in the nutrient-rich mud and provide above-water nesting sites for birds and underwater nesting sites for fish.

Available freshwater has always made human colonization of a territory easy; civilizations tend to spring up near freshwater locations. Lack of available freshwater has historically precluded human habitation. Certain aspects of human interaction have negative effects on the health of overall freshwater systems. In earlier times, dumping wastes into moving water seemed like a good idea because the current moved it away. As the practice magnified, only the first person upstream could claim unpolluted water. Today, attempts to minimize man-made disturbances are inviting a return to more pristine times.

Marine

Ocean biomes are divided into three zones, which show marked differences:

- Intertidal
- Neritic
- Pelagic

The *intertidal zone* is the shallow water where water meets land. It is usually tidal, meaning that the interface changes regularly, exposing land at low tide. Organisms inhabiting this area must adapt to hydration and dehydration, the impact of waves, and the return peril of ocean currents. As an example, sea stars have specialized appendages called tube feet to cling to rocks. The intertidal zone is also a favorite gathering point for seals, birds, and sea lions. It is also a favorite play area for humans.

The *neritic zone* is the most productive marine area because it is shallow enough for light penetration. This *photic zone* bustles with plankton and algae and boasts an extensive food web. It is the intersection between nutrient runoff from land and upwelling from the ocean bringing marine organic matter. In tropical areas, invertebrate corals create huge coral reefs as their limestone shells pile up in the shallow waters. Although only the outer layer of coral is actually living, the reef provides habitat for a plethora of aquatic life. Interestingly, the coral has a mutualistic relationship with algae; the coral provide an attachment site, and the algae supply products of photosynthesis. Like tundra areas, coral reefs regenerate slowly after an injury, such as an oil spill, random chipping of the coral for souvenirs, and gross coral removal for use as building materials.

The *pelagic zone* is the vast open ocean. At one time it was believed that the bounty of the ocean was endless. We now know that the pelagic zone is not very productive. Only the upper layer is photic; the bulk of the deep water is *aphotic*, meaning that only heterotrophs can live there. Occupants of these murky depths are largely scavengers, bottom feeders, decomposers, and interesting almost prehistoric-looking fish.

Animals at these depths experience tremendous water pressure and cold temperatures, so they have diminutive skeletons and a reduced metabolism. Certain bottom-dwelling fish have a mutual relationship with luminescent bacteria in the skin of their bodies. The bacteria give off light when the fish is excited. Although the photic zone receives adequate light, it is usually nutrient deficient, preventing protist production. Animals that live in this area are accustomed to traveling to access food sources. The open-ocean photic zone is home for aquatic mammals such as whales, cartilaginous fish such as sharks, and invertebrates such as squid.

Fortunately, the ocean is so large that man's interference with the laws of nature tends to be isolated. Although the *Valdez* oil spill created havoc for Alaska and sections of Canada, it had little effect on the South Pacific. The great variety of marine life is located in the photic zone in the shallow waters of the continental shelf.

The Least You Need to Know

- The earth is divided into nine terrestrial biomes and three aquatic biomes based on climate, abiotic, and biotic characteristics.
- Sometimes natural disasters are good for a biome.
- Plants and animals have developed unique structures and strategies that enable them to live in the biome.
- Biomes are fragile, some more than others; man is often a destructive force.
- Biomes change slowly over geologic time.

Chapter 24

Population Ecology

In This Chapter

- Biotic and abiotic factors affecting populations
- Existing methods and limitations to estimate the size of a population
- How population growth models respond to carrying capacity
- Human population growth rates and the logistics model

Ecology is the study of how living things interact with each other and their environment. In an ecosystem, everything affects everything else. Abiotic factors such as temperature, the amount of sunlight or rainfall, and the amount of pollutants in the air affect the biotic factors such as the growth and reproduction of individuals, species, *populations*, and *communities*. These abiotic factors then, in turn, affect the biotic factors, and so the cycle continues.

As predator-prey interactions, environmental pressures, and species-species activity occur, the density and dispersion of populations in an area fluctuate according to the rules of natural selection. Except for human populations, we may be in a growth stage that challenges the carrying capacity for this planet as more and more areas fail to provide for the needs of the current inhabitants. There are models to describe these fluctuations, and they are easy to understand. Simply look at the pictures and read the text; it's easy as that.

Population Properties

Population dynamics are affected by both biotic and abiotic factors. In determining the size or number of individuals in a population, density and dispersion as well as census techniques must be considered.

Density and Dispersion

The density of a population is how tightly the individuals are packed into a given area. For instance, at the recent millennium mark, in the United States the average human population density was 30 individuals for roughly every .38 square miles or per square kilometer. In Japan, it was 330 per .38 square miles or per square kilometer. Although useful, these numbers do not explain whether people are living in cities or evenly dispersed throughout the country. In studying populations, it is often important to know how they are dispersed.

There are three classifications of dispersion:

- **Clustered.** A clustered population distribution occurs when the individuals are living in groups, such as a herd of wild horses, school of fish, or humans living in a city.
- **Even.** An even distribution of individuals is the ordered placement of individuals, often because of spatial requirements. Farmers typically use even distribution when planting a crop because they know the requirements for that plant. Certain plants, such as a creosote bush, contain germination-inhibiting chemicals that prevent new plants from growing in their spatial-requirement zone. These plants also supplement this strategy with fibrous surface roots that generally absorb any moisture, depriving potential competitors. As a result, the creosote bushes are evenly distributed.
- **Random.** Random distribution is the opposite of even distribution in that there is no identifiable pattern to the dispersion. Randomness applies to the natural growth of most trees in a forest, nonschooling fish in a neritic zone, and middle school students at their first Friday-night dance.

Examine the illustration *Density patterns.*

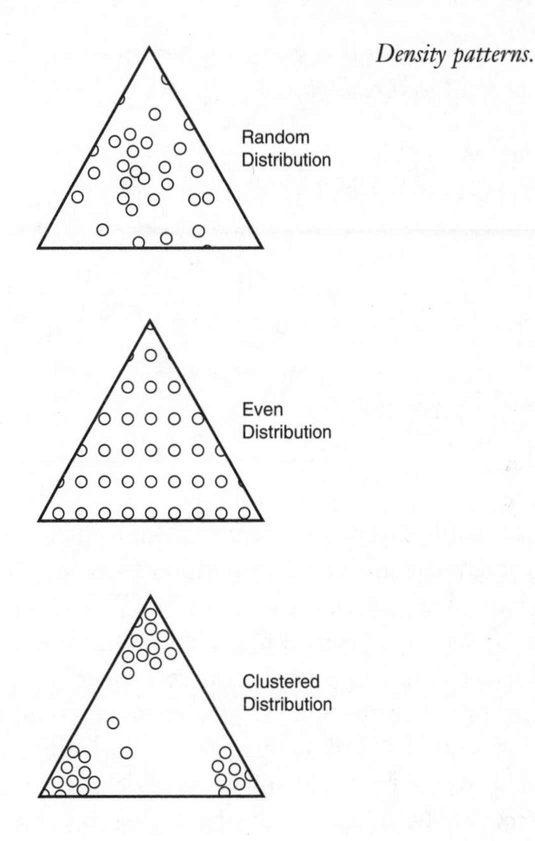

Density patterns.

Census Techniques

Before examining or experimenting on a population, it is important to have a reliable estimate of how many are actually there and their dispersion type. The most accurate accounting of the population is to count each individual in the defined area. However, if you are counting the number of grasshoppers in a 10-acre field or the number of poplar trees in a state forest, this technique may not be the preferred method because of the sheer numbers and the likelihood that a mobile animal will be counted more than once. Statisticians and scientists have collaborated to develop many sampling techniques that allow the researcher to approximate the population.

Count and multiply is a technique in which the ecologist counts the number of organisms in a finite area and then multiplies that number by the total area. For instance, there are 3 grasshoppers in a 1-acre section, and there are 10 sections, so $3 \times 10 = 30$ grasshoppers in that 10-acre area. Of course this number may or may not be accurate, but it does serve as a quick estimate. This technique assumes that all area-sections are homogenous and there is no reason to think that there might be more or fewer grasshoppers in a different section (in other words, an even distribution). Refer to the illustration *Count and multiply*.

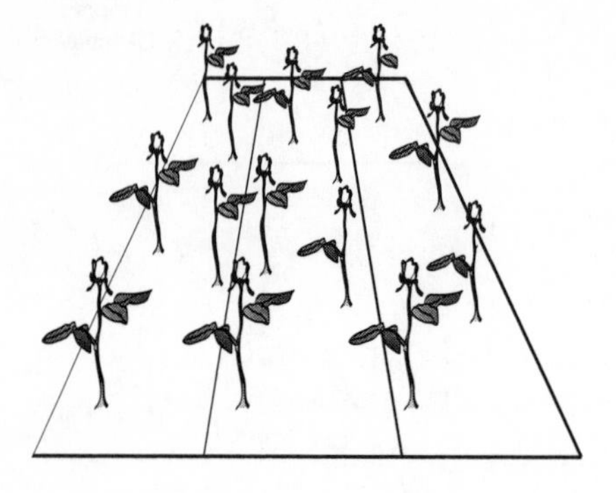

Count and multiply.

Line transect is a technique that is useful if the territory is heterogeneous and is not likely to contain an even distribution of species. Researchers use a predetermined length of a line, say 50 yards or meters. Then either starting in a random location or from a fixed-grid orientation, the line is laid out and all the targeted organisms in a fixed interval, such as 1 yard or meter, are counted on both sides of the line. So for a 50-yard or meter line and a 1 yard or meter counting distance on both sides (which equals a total of 2 yards or meters), for each transect of this type a 50 yards or meters × 2 yards or meters = 100-square-yard or meter area is sampled. In small areas, one transect may be enough. The assumption for a line transect is that the counted species are evenly or at least randomly distributed, because if they are clustered like cows in the shade, the results will be inaccurate. Refer to the illustration *Line transect.*

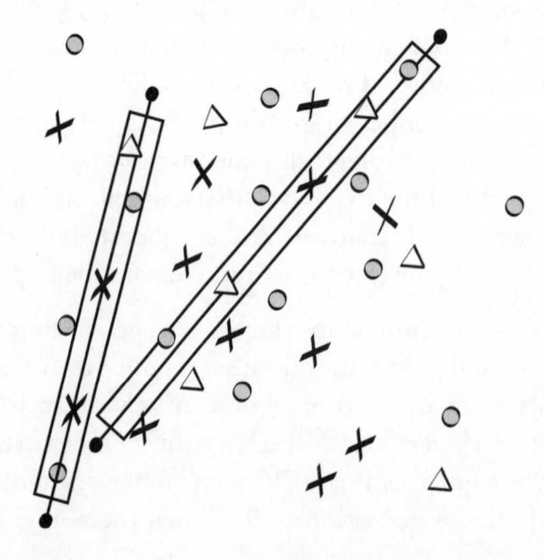

Line transect.

The *mark-recapture* method works for any population distribution. In this technique, the targeted organism is trapped, counted, marked, and released. The traps are reset and checked again at equal intervals until the termination of the experiment. The data is organized into the following mathematical relationships after each trapping to generate an estimate of the total population from this sampling technique. The estimated number = the number of marked individuals caught × total catch ÷ the recaptured marked individuals.

A lepidopteron, or one who studies butterflies, wanted to know how many swallowtail butterflies visited a particular butterfly bush each day, so he netted every swallowtail butterfly, counted it, and placed a small touch of fingernail polish on the back of the midsection and released them. He followed this procedure during the long summer daylight hours for two weeks. At the end of this time, he concluded after analyzing his data that the butterflies were local and apparently not migrators because the same butterflies were trapped repeatedly. It would be interesting to do a related study based on this idea to try the same procedure late in the summer to determine whether migratory butterflies stop for refueling.

Bionote

An interesting drawback to the mark-recapture method is the presence of "trap-happy" organisms that are caught repeatedly. Perhaps they are overly attracted by the bait, but their omnipresence can skew the data.

Population Growth

Sampling techniques are useful to determine the current number and distribution of a population, but they do not attempt to predict future growth patterns. Any prediction of future growth has several noticeable assumptions. First, an assumption about the birth and mortality rates: Are they constant, seasonal, or random? What about the effect of immigration and emigration in the future trends? The exponential growth model and the logistic model are two models that present graphical predictions based on these elements.

Exponential Growth Model

The *exponential growth model* assumes that the growing conditions are ideal and there is no limit to growth such as predation or limited resources. The resulting growth is rapid and becomes exponential after a few generations. For instance, most fruit lovers know that the fruit fly, *Drosophila melanogaster*, multiplies rapidly in the kitchen. If that growth is allowed to continue, assuming new fruit and no pesticides, the fruit flies would literally be everywhere.

Let's assume that under normal kitchen conditions your brand of fruit flies birth 10 offspring every 5 days. So after 5 days, there are 5 + 2 (parents) = 7 fruit flies. Approximately 5 days later, assuming half of the population is female, 4 females × 10 offspring + 3 males = 43. After 5 more days, the same scenario = 22 females × 10 offspring + males = 242 or so adults. Within a couple more generations, the numbers get large quick. In fact, their growth rate becomes *exponential.* Graphically, an exponential growth rate always looks like the letter *J* because the growth may start slow, but it increases rapidly. Refer to the graphic illustration *Exponential growth.* The exponential model describes the growth of a population as one that grows faster as it gets larger. In reality, there are limitations to exponential growth.

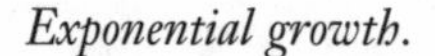

Exponential growth.

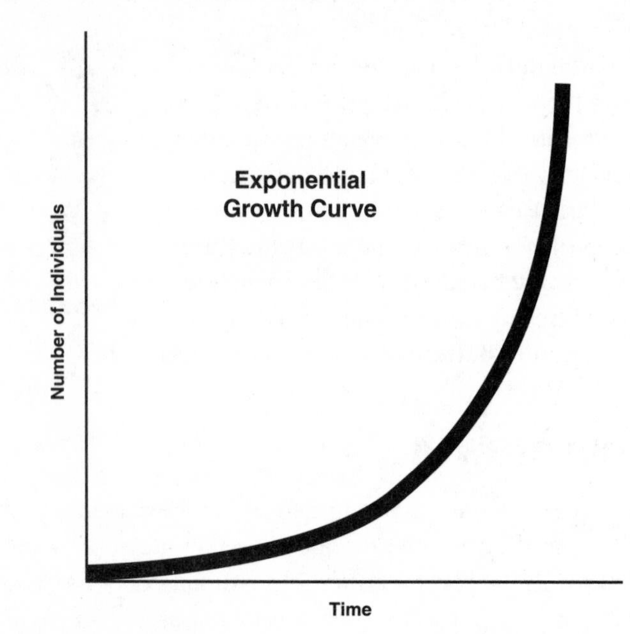

Limitation of Exponential Growth

In the natural state, population growth curves bend at a prescribed level because of a confluence of biotic and abiotic factors such as increased predation and dwindling food allotments per individual. *Limiting factors* define growth rate and establish a relatively stable population size for a particular environmental area and a given population, called the *carrying capacity.* The carrying capacity for a prescribed territory is the maximum number of individuals that a given environment can sustain. It is established by a combination of limiting factors that may or may not be related to the population properties:

- Density dependent
- Density independent

Density-dependent limiting factors are those limitations that operate because of close proximity of individuals in a population; so a densely populated area is more affected than a sparsely populated territory. There are two main types:

- Competition
- Predation

Competition increases as density increases because more individuals are in need of a static or finite amount of resources such as food, water, and habitat. The force of competition leads individuals and species to develop either new ways to get their necessities (relocating, for example) or minimizing the need for the resource (water conservation, for example). New techniques may require new habitats and structures, which may create more diversity.

Successful predation increases because the number of available prey in a given area also increases. Increased successful predation creates reproductive success for the predators and decreasing numbers for the prey. This continues until the number of predators becomes too high and they catch enough prey so that the prey becomes scarce. The predator population then decreases, which in turn increases the prey population. Their growth rates are often intertwined. Predation forces both the prey and predator to adapt increasingly sophisticated structures and strategies (see Chapter 14).

Density-independent limiting factors operate to reduce the population regardless of the number of individuals in the population or how they are dispersed. Three main types of density-independent factors affect all natural populations:

- Weather and climate
- Natural disasters
- Human interaction

Weather acts equally on all populations but may favor one over another. As an example, warm temperatures in the spring favor population growth in insects as they hatch, but a late-spring freeze may eliminate the newborn population, decreasing the population growth. Likewise, prudent fruit growers know to plant fruit trees in areas that receive cold spring breezes, usually off of water, to prevent early budding in the trees and increase growing time in late summer as warm breezes protect the plants from freezing. Native plants and animals are normally well adapted to seasonal changes, but organisms thriving in fringe areas or caught unprotected in unusual weather patterns may have precipitous die-offs.

Natural disasters may also help some species while harming others regardless of their size. Fires are frequent in chaparral biomes and are costly to plants and animals not well adapted for the disaster. Populations of these organisms, such as invasive plant species, may be brought to zero, while indigenous plants reap the benefit of reduced competition and the addition of wood ash as a soil amendment. Oddly, the hot and dry summer months in the Mid-Atlantic states are also the months of their greatest rainfall totals. This phenomenon is true not because of typical weather patterns, but in response to the occasional hurricane or tropical storm that dumps torrential amounts of rain in a short period of time. The drought that lasted from 2000 through 2003 was officially over when Hurricane Isabelle roared through and brought rain in buckets for the entire area! The drought conditions favored warm-water aquatic life such as the Maryland blue crab, but the decreased cool water flow in mountain streams was a limiting factor for brown trout. When the rains came, cool oxygenated water helped the trout recover. The massive influx of fresh water sent the crabs to more saline areas. The increased oxygenated water helped both types of aquatic life.

Often innocent human interaction spells doom for some species and promotes growth in others. Urban sprawl has eliminated habitat for certain species such as black bears but has increased habitat for songbirds, raccoons, and deer. It has also decreased natural predators such as snakes and eagles, which then increased the rat and mice population. Typical homeowner landscaping with a yard of grass and a bed of flowers often replaces the nature arboreal or grassland species. Introduced ornamentals change the landscape for the original inhabitants as well. For every action, whether man-made or natural, there is a reaction or adaptation.

Logistic Model

The logistic model represents a more realistic analysis of population growth because it predicts growth based on predictable population-growth events as detailed by density-dependent and density-independent factors. When combined, these factors determine the rate at which a population will expand and ultimately determine the carrying capacity for that area. The logistic model assumes that a population will continue to grow exponentially until the environmental factors alter that rate. The stages that all populations go through to reach the steady state for a given species in a given area can be analyzed in three stages:

Stage 1. Exponential growth. The ratio of births:deaths is large.

Stage 2. Density-dependent and density-independent factors reduce the ratio of births:deaths.

Stage 3. Birth rates and death rates fluctuate but average 1:1.

Stage 1. Exponential growth as explained in the previous section is the absolute maximum amount that a population can grow with unlimited resources and no limiting factors.

Stages 2 and 3. Both density-dependent and independent factors subtract from the total population.

As competition, predation, climate, and disasters (both natural and human) take their toll on the population and the population resources, the graph of the steady state contains a continuous range of peaks and valleys, but the overall average detects no measurable increase or decrease. The steady state has been achieved, and it is equal to the carrying capacity for that area.

Close examination of the steady state often reveals sheer drops in a population followed by exponential growth. For example, a herbicide may eliminate all undesirable weeds except for one that is immune, which then reproduces in monster numbers. Likewise, one predator may virtually eliminate prey, which then starves the predator. Both rebound with sharp inclines. Careful analysis of the steady state for a species or territory must include the element of time to provide a picture of the steady-state history. Complacency in certain areas regarding the cost of human population growth may be the result of minimizing the time element. The human population is experiencing a Stage 1 growth spurt; environmentalists see Stage 2 and Stage 3 factors on the horizon.

Human Population

Homo sapiens began as nomadic *hunter-gatherer* tribes or clans that moved periodically to greener pastures. The population of these societies remained low and was controlled substantially by density-independent factors that directly influenced their food supply. Approximately 10,000 to 12,000 years ago, *Homo sapiens* adopted a domesticated lifestyle that was based on solid agriculture and animal husbandry principles, so the food sources were more reliable and under greater control. This advancement reduced those density-independent factors, which increased the carrying capacity for that area, and the human population expanded accordingly.

Following several rounds of the Black Death (or bubonic plague), which killed an estimated 25 percent of the European and Mediterranean population, the industrial age spawned the latest increase in the human population. The Industrial Revolution freed some people to devote more time to education and learning, which led to improvements in health, sanitation, and medicine. Again, these advancements have minimized certain density-independent factors, and the resultant human population growth index clearly shows a Stage 1 growth pattern. It is assumed that there is a finite carrying capacity for humans on Earth. We just don't know how far our technology can continue to increase the carrying capacity. The future belongs to tomorrow, but it may have been born yesterday.

Curiously, industrialized countries have voluntarily decreased their birth rate so that they are doubling their population in excess of 100+ years or more. This stabilization may be caused by better contraception and delayed pregnancy. A female reproducing in her teens may have grandchildren when a female in her 30s may only have young children. With each generation, the gap gets wider. Underdeveloped countries maintain extremely high birth rates, which often intersect with territories that appear to be approaching their human carrying capacity. The countries least capable of handling a population increase are actually increasing the fastest! At some point, population pressures may force a political solution.

The Least You Need to Know

- Populations are regulated by density-dependent and density-independent factors.
- There are several techniques for estimating the size of a population, and they all have limitations.
- The exponential and logistic models explain population growth in graphic formats.
- Human and natural interventions affect some species more than others, and they affect the overall population.
- The human population is in a Stage 1 growth pattern, in which the ratio of births:deaths is large.

Chapter 25

Succession and Energy Transfer in an Ecosystem

In This Chapter

- How ecosystems change over time
- The ecological progression toward a climax community
- Energy flow through an ecosystem
- Obstacles to the natural recycling of biologically important substances
- Effects of pollutants on chemical cycles

As a young lad, I was forced to mow the yard way too many times. I often wondered what it would be like to let it go; just let it grow. This chapter looks at succession as an ordered change in the plant and animal life in an area based on the most fit surviving and changing the environment until the ultimate most fit arrives and assumes domination (that is, until the next disaster restarts the succession clock over again).

The energy flow through any ecosystem travels from the ultimate source, the sun, to autotrophs (plants), and then to heterotrophs (animals). Although food webs can be quite involved, the flow of energy always travels the same way. Humans are at the top of the pyramid because there aren't many animals that prey upon us for food. Most humans are either too tough or too skinny to make a good meal anyway. In pyramids that do not involve humans, other animals are at the top.

This energy flows in one direction through an ecosystem, but most inorganic nutrients flow in a cycle so they can be reused. The four most biologically important cycles are examined in detail, with great pictures to brighten up the page. They are easy to remember because they relate to you in everyday life.

Succession

It is reasonable to assume that populations change as environmental factors affect their life cycle. Is it also safe to assume that the environment changes? It was once believed that ecosystems were stable in that they did not change much. It is now thought that change within an ecosystem is both constant and often cataclysmic. Catastrophic events such as a wildfires, hurricanes, or volcanic eruptions create disturbances within an ecosystem on a periodic basis. Whenever an event of this magnitude occurs, it often decimates the plant and animal populations living there.

When Mount St. Helens erupted in 1980, the ground for miles in any direction was devoid of life. The massive wildfire in Yellowstone National Park in 1988 burned more than one third of the park, leaving only scorched earth. Interestingly, pictures taken today of the same barren landscape show vegetation and animal life inhabiting both areas. The process of plants and animals colonizing or recolonizing an area is called *ecological succession*. There are two types of ecological succession:

- Primary
- Secondary

Primary succession occurs when the colonization occurs in areas where no previous life inhabited, such as Mount St. Helens after the lava overflow. Primary succession normally occurs in areas with no soil, such as a volcanic lava field or a city sidewalk. Typically the first colonizers or *pioneer species* are algae, lichens, and mosses that can attach directly to a rock surface. Sometimes those are all of the life-forms that can establish and grow. At other times, the pioneer species trap airborne particles, die, and decompose, and thereby create modest pockets of soil over time. Sometimes that is enough for larger species, perhaps a small bush, to become established. The presence of the bush may create more soil, and the roots may assist the mechanical and chemical weathering of the rock to create small cracks or holes that also trap water and debris. Larger plants become established and what was once a barren lava field becomes a coniferous forest, in the case of Mount St. Helens, or a tropical forest as on the slopes of Kilauea in Hawaii.

Secondary succession is more common because it is the recolonizing of an area that still contains soil after a disastrous event, such as a wildfire that has vanquished the life-forms. The intense Yellowstone fires denuded the area of life; however, the soil remained intact so that seeds already present in the soil and those brought in by wind and other agents germinated, grew, and re-created a vegetative landscape. Secondary succession is faster than primary succession.

Ecological Succession

Ecological succession is the replacement of one species by another species more fit for that environment. Typically the process begins with plants, and animals follow. As a new plant species grows, it almost always changes the environment, favoring one species over another.

In certain areas, pine trees colonize faster than deciduous hardwoods, so expansive pine forests dominate secondary succession areas after a fire or when a crop field is abandoned. The pine trees shade the ground so that certain plants that need direct light fail to grow and reproduce. Thus the pine trees change the environment to exclude light-loving plants. Maple trees grow well in partial shade and are able to grow in a developing pine forest. As the maple trees grow, their expansive broadleaf structure provides even more shade, which in turn may be enough to prevent young pine trees from growing. Over the course of years, maple trees slowly replace the pine forest. Sometimes if wind or lightning destroys a pine tree, the opening is quickly filled either with more pine trees, or grasses, shrubs, or sun-loving trees such as oak. Regardless, in temperate areas, deciduous hardwood trees are generally the *climax community*. In colder areas, such as taiga, deciduous hardwoods are not as well adapted, so coniferous forests are the climax community.

Bioterms

A **climax community** is one where the plant and animal populations in an area cannot be replaced by more fit species without an environmental disturbance.

The simplest model of secondary succession is the periodic disaster that regenerates the temperate grassland:

- Periodic fires burn off all aboveground vegetation.
- Grasses regrow faster than woody plants and prevent them from becoming established.

- Grazing animals eat grasses, which regrow, and the meristimatic area of surviving woody plants, which do not regrow.
- Grasses become the dominant species. Fires burn off the area, and the cycle is renewed.

As the plant scenery changes, the opportunity for animals to occupy new territories also increases. In primary succession, only invertebrates are present until more substantial foliage develops. In tundra areas, however, grazing herbivores readily consume pioneer plants as a staple to their diet. Grazing herbivores always attract a predatory crowd, so even small pioneer plant offerings can produce a food chain.

As the plant species become more advanced and diverse, the animals that inhabit them follow suit. As beach areas become recolonized after a hurricane, the pioneer beach grasses give way to woody shrubs that in turn are overtaken by woody trees. In a parallel succession, advanced invertebrates such as arthropods are replaced by small vertebrates, such as field mice, which are then preyed upon by hawks and snakes. Dominate species of plant and animal life change as they both succeed themselves toward a climax community. A disturbance restarts the succession clock, perhaps back to the pioneer species depending on the severity of the event.

Energy Flow in an Ecosystem

Even when an ecosystem is in the dynamic process of succession, energy transfer occurs in patterns similar to the climax community. Energy flows in and out of an ecosystem and is not contained exclusively within an ecosystem, such as the primary source of energy, the sun. Typically energy is transferred and transformed in a series of exchanges that provide usable energy for the living organisms in that area.

Food Chain

The movement of energy can be explained by an analysis of a typical *food chain*. A food chain discloses the "who eats whom" in a system. The following is an example of a simple food chain:

Radiant energy from the sun → plant → herbivore → carnivore → decomposer

The solar energy of the sun is converted into the chemical energy of glucose in the plant by photosynthesis. Some of the glucose is used by the plant for respiration; the rest is stored as starch. The glucose, starch, and other organic compounds are consumed by herbivores that convert some of the chemical energy into heat (ectotherm) and the remainder of the energy into usable energy for normal body function, while the leftover energy is converted into stored chemical energy, glycogen or fat. Likewise, the carnivore converts the captured chemical energy of the herbivore into heat energy, usable energy, and also stores the leftover as glycogen. Finally, the organic remains of the carnivore are recycled by the decomposers back into inorganic molecules that are usable by plants, and the cycle begins again.

Using the same simple illustration, we can specifically identify organisms to replace the generic types found in the original template. It is also important to note that at each new stage, called *trophic level,* the amount of energy transferred is only 10 percent of that contained in the previous organism:

sun → grass → rabbit → fox → bacteria

If we assume that hypothetically the sun provided an original input of 100 *kilocalories* (kcal) of energy into the grass, then only 10 kcal would be transferred to the rabbit, 1.0 kcal to the fox, with only 0.1 kcal of the original energy absorbed by the plant available to the bacteria.

Given that the first law of thermodynamics states that energy cannot be created or destroyed, where did it go as it decreased by roughly 90 percent between each *trophic level?* The answer is the second law of thermodynamics, which states that energy can by interconverted. In the process of energy conversion, most of the energy is wasted as heat energy that is absorbed by the environment.

Bioterms

A **kilocalorie** is the amount of heat needed to raise the temperature of 1 kilogram of water by 1 degree Celsius. A **trophic level** is the feeding structure in a system. Each feeding level is a trophic level. They are helpful in monitoring energy flow and chemical cycling in an ecosystem.

Food Web

Simple food chains seldom exist in nature because of the diverse and overlapping lifestyle of organisms in a given area. Normally energy transfer is analyzed in *food webs* that are more inclusive than food chains. Refer to the illustration *Estuaries food web,* which shows a simplistic estuarine food web.

The bay grasses receive their solar allotment, which they convert into energy and organic compounds. They are consumed by crabs, ducks, and fish. The energy transferred to these primary consumers is 10 percent of the original amount contained by the grasses. Great blue herons, humans, and hawks are secondary consumers receiving 10 percent of the original 10 percent of the starting energy. Fish and crabs prey upon each other at various times during their life cycle, making them secondary consumers as well. A food web sometimes contains a tertiary consumer, but seldom contains a quaternary consumer.

Estuaries food web.

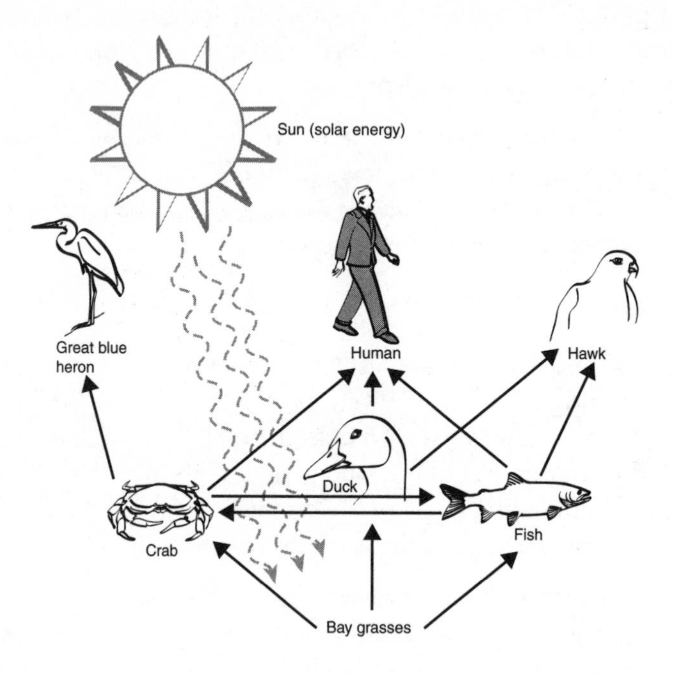

Chemical Cycles

Chemical transfers within an ecosystem are as important as energy transfers. Chemical transfers are different from energy transfers because the chemicals tend to stay within the ecosystem. The cyclic chemical systems interact between the biotic and abiotic components of an ecosystem. Four chemicals are of primary importance to biological systems: water, carbon, nitrogen, and phosphorus.

Water Cycle

The water cycle hasn't changed much since you learned about it in elementary school; however our understanding of the three physical processes that drive the cycle has improved.

Water is evaporated from the environment or transpired from plants and is released into the atmosphere as the invisible gas, water vapor. Water vapor is cooled, causing it to lose the kinetic energy of movement, thus pulling individual molecules closer together and finally into droplets. As the droplets coalesce, grow, and become heavier, gravity pulls them to the earth's surface. This is the ultimate source of all natural fresh water, which is readily usable by plants and animals.

Water is either used by living organisms, stored underground, or captured above ground in rivers, lakes, and oceans where evaporation begins the cycle by energizing individual water molecules to a change of state from liquid into the gas, water vapor. Refer to the *Water cycle* illustration.

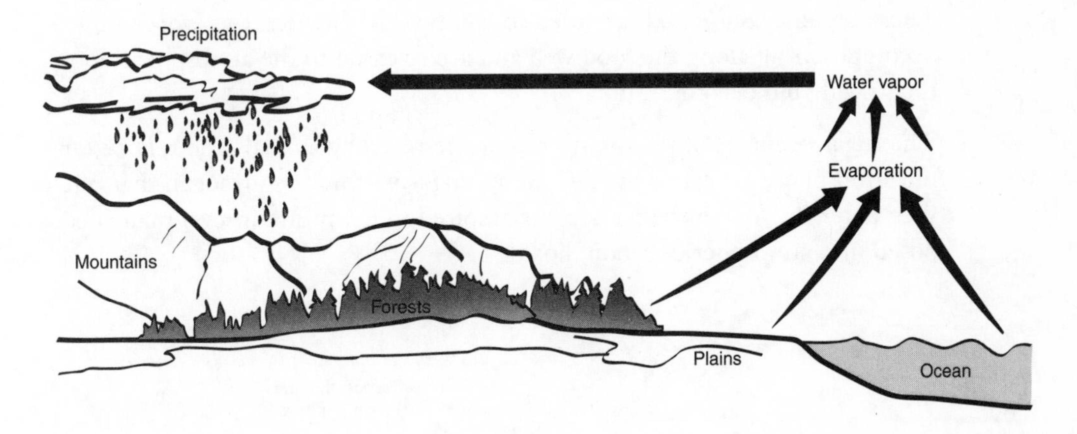

Water cycle.

The water cycle may be the next eco-disaster. Already we know that unhealthy chemicals dissolve in atmospheric water and return to Earth (in the form of acid rain, for instance). The decreased transpiration from denuded tropical rainforest areas reduces the amount of water vapor released to the atmosphere. Atmospheric ozone holes may allow greater ultraviolet light to penetrate the earth's atmosphere and melt polar and mountain ice, releasing more water vapor into the air. Reclaiming desert areas requires a substantial drain of subsurface water. These underground aquifers do not replenish as fast as they are depleted, so other options eventually must surface.

Carbon Cycle

Carbon is the element that creates organic molecules, which compose all life forms. It is especially important that useful carbon sources are available for all living things. Refer to the illustration *Carbon cycle.*

The simplest carbon cycle is the direct transfer of carbon in the form of carbon dioxide as a waste product of respiration to a green plant to be used as a raw material for photosynthesis, which generates the waste product oxygen needed for respiration.

An interesting self-perpetuating ecosystem can be established in a closed container with aquatic plants and herbivores. The waste products of one organism are required by the other, and vice versa. Earth has often been referred to as a closed container.

The carbon cycle has three steps:

1. Atmospheric carbon dioxide is transformed during photosynthesis into a sugar molecule and other organic compounds.
2. These organic compounds are used to compose herbivores, carnivores, and decomposers all along the food web and are released to the atmosphere upon their death and decomposition.
3. The sugar products of photosynthesis are respired by animals and released into the atmosphere as carbon dioxide. Some sugar is stored as glycogen that can then enter the food web and become respired in the future or eventually decomposed into atmospheric carbon dioxide.

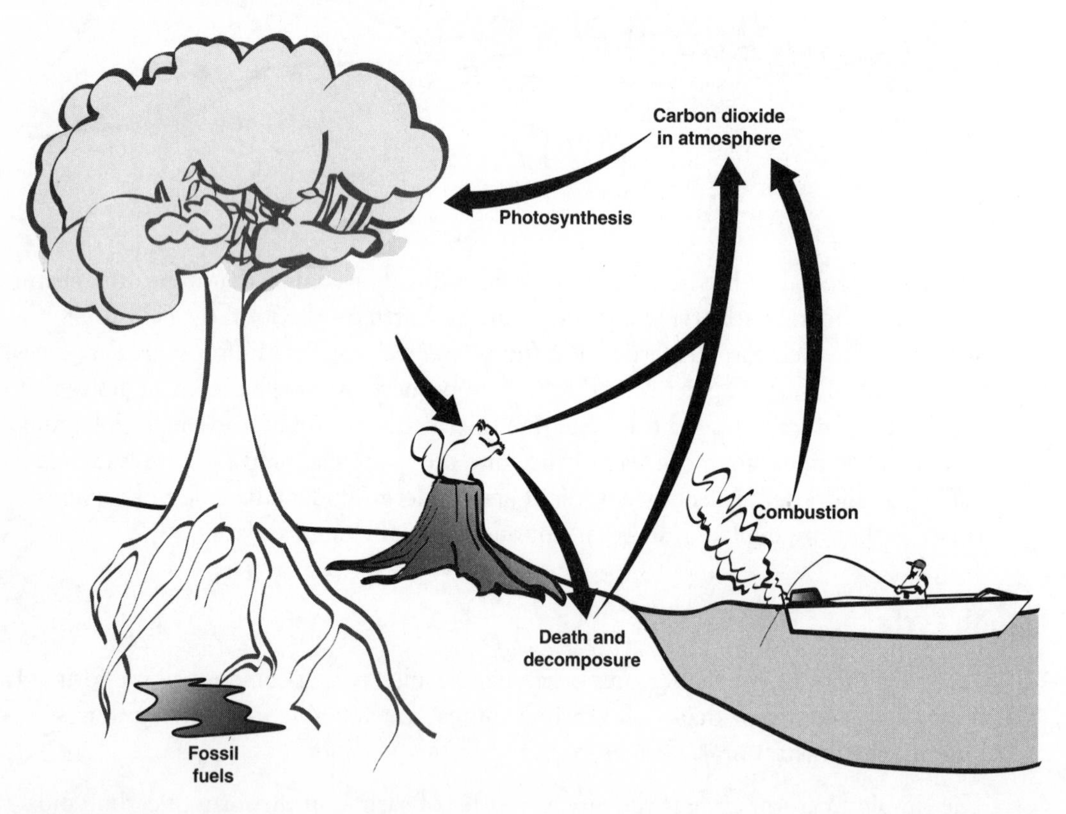

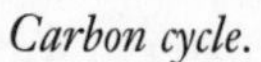
Carbon cycle.

Natural cycling of carbon and carbon products has been intensified by the increased burning of fossil fuels such as coal and oil. It is true that one of their waste products is carbon dioxide, which has an existing natural recycling procedure. It is also true that the released carbon dioxide is adding to the overall carbon dioxide concentration in the atmosphere because the fossil fuels are all subterranean. Burning fossil fuels liberates carbon dioxide from underground deposits and adds it to the global ecosystem. Environmentalists theorize that additional carbon dioxide, a greenhouse gas, in the atmosphere may promote global warming.

Nitrogen Cycle

Nitrogen constitutes about 78 percent of Earth's atmosphere, but exists in an unusable form. Oddly enough, the atmospheric nitrogen (N_2) is converted into a form usable by plants (NO_3 or NH_4) by the action of subterranean bacteria. This process is explained in four steps, as shown in the illustration *Nitrogen cycle.*

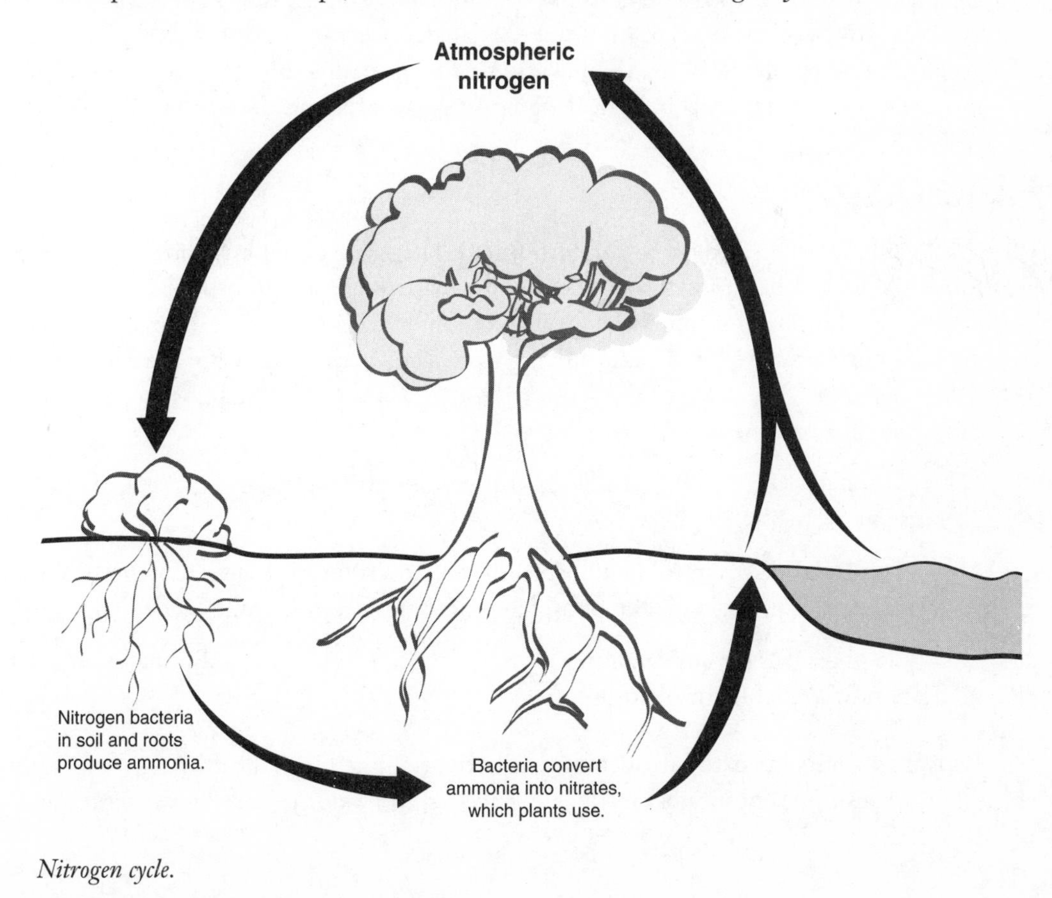
Atmospheric nitrogen
Nitrogen bacteria in soil and roots produce ammonia.
Bacteria convert ammonia into nitrates, which plants use.

Nitrogen cycle.

Nitrogen-fixing bacteria in the soil and in the legumes of some plants, such as soybeans, or cyanobacteria in aquatic systems, convert atmospheric nitrogen (N_2) into ammonia (NH_3), which absorbs hydrogen from water to become an ammonium ion (NH_4^+). Nitrifying bacteria convert ammonium (NH_4^+) to the usable nitrate (NO_3^-) for plant usage. The plants die and the nitrogen is released to the soil as ammonium (NH_4^+) by decomposers or they are consumed by herbivores, which enter the food web, and eventually the decomposers also convert their nitrogen into ammonium. The ammonium is then available for conversion into nitrates for reuse. Denitrifying bacteria convert nitrates in the soil back to atmosphere nitrogen (N_2), making it unusable to plants, but completing the cycle.

Normal nitrogen loads are kept in balance by soil bacteria. Some environmentalists fear that the desire for green suburban lawns may overload the natural cycle with excess nitrogen. In some areas, huge nitrogen runoff from lawns and farms has loaded the waterways with usable nitrogen that produces an unnatural algal bloom. When the algae die, their decomposition consumes all available oxygen, thus suffocating fish and other aquatic aerobes. Nitrates in water partially convert into nitrites (NO_2^{-2}), which are human metabolic poisons. Wastewater treatment plants do not always remove the nitrogen, which then spills into local waterways, exacerbating the problem.

Phosphorus Cycle

Water, carbon, and nitrogen are all airborne during their cycle, but phosphorus has a terrestrial base. The original source of all phosphorus is rock. Phosphorus is almost always found in living systems chemically combined with oxygen to form a phosphate group. Phosphates are important to living organisms because they are the high-energy components of ATP. Refer to the illustration *Phosphorus cycle*. The phosphorus cycle can be summarized in three steps:

1. Phosphates (PO_4^{-2}) are weathered from existing rock, making them available for use by plants.
2. The plants incorporate the phosphates and are in turn consumed by herbivores or they die. Either way, the phosphates are released eventually by decomposition.
3. The phosphates from decomposition and those released by additional weathering are now available for plant use.

Phosphates are similar to nitrogen in that both are linked to unhealthy algal blooms. They are also present in most fertilizers, and in some water released after wastewater treatment.

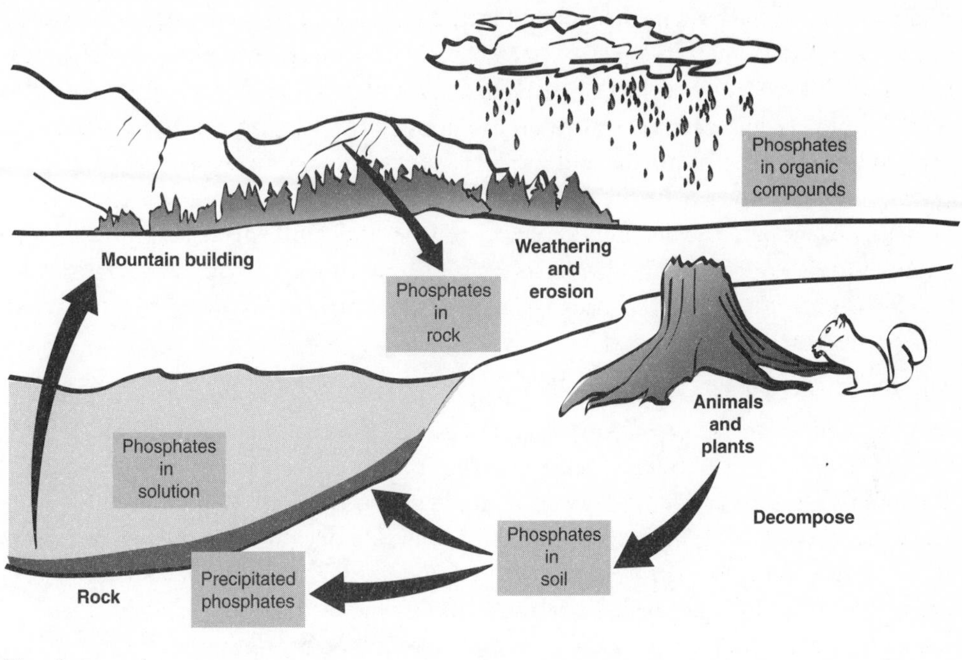

Phosphorus cycle.

The Least You Need to Know

- Succession is an ordered process that natural systems use to advance or restore ecosystems back to their climax community.
- There are two types of succession (primary and secondary) that proceed at different speeds and define the sensitivity of our environment.
- Energy flow through an ecosystem can be pictorially represented in food chains and food webs.
- At each new trophic level, only 10 percent of the available energy contained in the previous organism is transferred to the next trophic level.
- Water, carbon, and nitrogen are biologically important chemicals that are recycled using the atmosphere as a medium of transfer; phosphorus is a terrestrial recycling event.

Glossary

abscisic acid A plant hormone that inhibits cell division and promotes dormancy; also interacts with gibberellins in regulating seed germination.

acquired traits According to Lamarck, they were characteristics that an organism developed during their lifetime that could be transferred to the offspring.

Acrasiomycota The phylum containing cellular slime molds.

adaptation An anatomical structure, physiological process, or behavioral trait that improves an organism's likelihood of survival and reproduction.

adaptive radiation process Also known as divergent evolution, in which one species gave rise to many species that appear different but are similar internally.

ADP (adenosine diphosphate) A lower-energy molecule involved in the energy metabolism of a cell; formed when ATP loses a phosphate group and the related energy. See *ATP*.

aerobic Requiring molecular oxygen to metabolize.

allele A frequency measurement that determines how frequently the allele expression of a particular gene arises in a population.

allotrophic speciation The development of a new species because of geographic separation.

alpha helix The spiral shape resulting from the coiling of a polypeptide.

alternation of generations The switching back and forth between the production of diploid and haploid cells, which is common in plants.

amine An organic compound with one or more amino groups.

amino acids An organic molecule containing a carboxyl group and an amino group that serves as a monomer of proteins.

analogous structures Structures that are similar in appearance and function but have different origins and usually different internal structures.

aphotic zone The region of an aquatic ecosystem beneath the photic zone, where light does not penetrate enough for photosynthesis to take place.

artificial selection Selective breeding of domesticated plants and animals to promote the occurrence of desirable inherited traits in offspring.

atom The smallest unit of matter that retains the properties of an element.

ATP (adenosine triphosphate) A high-energy molecule that serves as an energy source for all living cells; formed when a phosphate group and accompanying energy is added to ADP. See *ADP*.

autosome Types of chromosome code for all body features except for sex cells.

auxin A plant hormone that regulates cell elongation; unequal auxin concentration causes bending of a plant stem.

bacteriophage A virus that infects bacteria, also called a phage.

benthic zone The bottom of a freshwater lake, pond, river, stream, or ocean floor.

binary fission A means of asexual reproduction in which a parent organism divides into two individuals of about equal size.

biological species concept A species whose members have the potential in nature to interbreed and produce fertile offspring.

biome A terrestrial ecosystem, largely determined by climate, usually classified according to the predominant vegetation, and characterized by organisms adapted to the particular environments.

budding A means of asexual reproduction whereby a new individual developed from a "bud" or outgrowth of a parent splits off and lives independently.

Calvin cycle A process in which a series of biochemical reactions convert carbon dioxide into a carbohydrate.

capsid A protein sheath that surrounds the nucleic acid core in a virus.

carbohydrates A class of biological molecules consisting of simple single-monomer sugars (monosaccharides), two-monomer sugars (disaccharide), and other multiunit sugars (polysaccharides).

carbon fixation The incorporation of carbon from atmospheric carbon dioxide into the useable carbon found in organic compounds.

carboxylic acid An organic compound containing a carboxyl group.

carrier protein A transport protein that carries a specific substance across a cell membrane.

Casparian strip A waxy barrier in the walls of endodermal cells in a root that prevents water and ions from entering the xylem without crossing one or more selectively permeable cell membranes.

catalyst A substance that speeds up the rate of chemical reaction without being changed or used up by the reaction.

cell The fundamental unit of living matter separated from its environment by a cell membrane.

cell membrane A cell structure that regulates the passage of materials between the cell and its environment and aids in the protection and support of the cell.

cell wall A cell structure found only in plants, algae, and certain bacteria that surrounds the cell membrane for protection and support.

cellular respiration The energy-releasing chemical breakdown of food molecules into a form that cells can use to perform work, composed of the following processes: glycolysis, the Kreb's cycle, the electron transport chain.

centromere region A region on the chromosome where the two sister chromatids are held together; also the site where spindle fibers attach to the chromosomes.

chaparral A biome dominated by shrubs adapted to periodic drought and fires, usually located where cold ocean currents circulate offshore to create mild, rainy winters and long, hot, dry summers.

chemiautotrophic Oxidizing an inorganic compound as a source of energy.

chemiosmosis The production of ATP using the energy of a hydrogen ion gradient across membranes to phosphroylate ADP.

chitin A tough carbohydrate found in the exoskeleton of all arthropods.

chloroplasts The photosynthetic organelle found in plants and photosynthetic protists, enclosed by two concentric membranes.

chromatin The combination of DNA and proteins that constitute eukaryotic chromosomes.

chrysophyta The phylum containing yellow-green algae, golden-brown algae, and diatoms.

cilia Specialized structures that serve in free unicellular organisms to produce locomotion, or in higher forms to produce a current to move substances.

ciliate A protist that has many hairlike structures (cilia) that aid in movement.

cladistics phylogenetic The method in which relationships are inferred and organisms classified based on the presence of derived character.

clone A genetically identical copy of a cell or organism.

codominance The shared expression of two different alleles of a gene in a heterozygote.

codons A group of three-nucleotide base sequences in an mRNA molecule that specifies a particular amino acid or polypeptide termination signal.

coenzymes An organic molecule (usually a vitamin) that acts as a cofactor to help an enzyme catalyze a reaction.

cohesion The attraction between similar molecules.

collenchyma cells A cell with a thick primary wall and no secondary wall, functioning mainly to support the growing parts of plants.

continental drift The common name for movement of the continents over geologic time, caused by plate tectonics.

convergent evolution The process by which unrelated species living in the same area become structurally similar as they adapt to similar environmental pressures.

covalent bond An attraction between atoms that share outer-shell electrons.

cristae A fold of the inner membrane of a mitochondrion.

crossing over The exchange of reciprocal segments of DNA by homologous chromosomes at the beginning of meiosis.

cytochromes Any of several cellular pigments that function in the electron transport chain as electron carriers.

cytokinesis The division of the cytoplasm to form two separate daughter cells.

Cytokenisis usually occurs during the telophase of mitosis, and the two processes make up the mitotic phase of the cell cycle.

cytoplasm The semifluid substance and everything inside of a cell between the plasma membrane and the nucleus.

dark reaction Also known as the Calvin cycle; reactions of photosynthesis that do not require light but use energy produced and stored during light reactions to make glucose.

dehydration synthesis A common chemical process also called a condensation reaction in which a polymer forms as monomers are linked by the removal of water molecules; may produce other molecules.

deletion The absence of a section of genetic material, which may cause serious mutations.

desertification The spreading of desert usually caused by overpopulation, climate change, and poor land management.

dichotomous key A method for the identification of organisms based on a series of choices between alternative characters.

diffusion The movement of particles—from an area of high concentration to an area of lower concentration.

dihybrid cross A mating that involves two sets of characteristics or alleles.

diploid The term used to indicate a cell containing two sets of chromosomes (2n), one set inherited from each parent.

directional selection Natural selection that acts against the relatively rare individuals at one end of a phenotypic range.

divergent evolution Also known as adaptive radiation, in which one species gives rise to many species that appear different but are similar internally.

DNA ligase An enzyme, essential for DNA replication, that catalyzes the covalent bonding of adjacent DNA nucleotides, also used in genetic engineering to join a specific piece of DNA containing a desirable gene of interest into a bacterial plasmid or other vector.

dominant A genetic trait that is expressed when its allele is homozygous or heterozygous.

duplication An error in meiosis in which the genetic material on a part of a chromosome is repeated; causes mutation.

ecosystem A territory encompassing all of the biotic and abiotic factors in that area.

electron transport chain A series of electron carrier molecules that transfer electrons to systematically release energy used to make ATP; located in the inner membrane of mitochondria, the thylakoid membranes of chloroplasts, and the plasma membranes of prokaryotes.

endocytosis The movement of materials into the cytoplasm of a cell via membranous vesicles or vacuoles.

endospore A dormant bacterial cell enclosed by a tough coating that is resistant to environmental stress.

endosymbiosisa A process of successful prokaryotic invasions from which the mitochondria and chloroplasts of eukaryotic cells probably evolved.

enhancer A nucleotide sequence that increases the rate of genetic transcription by increasing the activity of the nearest promoter on the same DNA molecule.

enzyme A protein that serves as a biological catalyst by increasing the rate of a chemical reaction without itself being changed in the process.

epoch An interval of time in a geologic time scale.

era The largest interval of time in a geologic time scale.

estivation An animal's state of reduced activity (torpor) during periods of environmental stress.

estuary A highly productive area where freshwater mixes with seawater.

eukaryote An organism whose cells have a nucleus enclosed by a membrane and are considered more advanced than prokaryotes.

evolution A genetic change in a population or species over generations caused by natural selection and mutation.

exocytosis The exit of materials out of the cytoplasm of a cell via membranous vesicles or vacuoles.

exon The sequence of nucleotides on a gene that gets transcribed and translated; it is also the coding portion of a gene.

extinct The term used to indicate species that have disappeared permanently and no longer contribute their genetic complement.

F2 generation The offspring from crosses among individuals of the F1 generation.

facilitated diffusion The passage of a substance across a biological membrane with its concentration gradient, aided by specific transport proteins.

fermentation An anaerobic process of cellular respiration that recycles NAD+ needed to continue glycolysis; also produces ethanol.

flagellum The threadlike structure that grows out of a cell and enables it to move.

fragmentation A means of asexual reproduction whereby a single parent can break off parts that regenerate into whole new individuals.

fragmentation effect See *fragmentation.*

gametogenesis The production of gametes.

gametophyte A haploid phase that produces gametes in the life cycle known as alternation of generations.

gene library All of the genes in a particular system.

gene migration The movement of individuals into or out of a population, which affects genetic diversity.

genes A discrete unit of hereditary information consisting of a specific nucleotide sequence in DNA (or RNA, in some viruses).

genetic drift A change in the gene pool of a population due to chance.

genetic equilibrium The stable state in which conditions do not favor extreme phenotypes.

genetic recombination The use of genetic engineering techniques to alter the DNA of an organism.

genotype The genetic makeup of an organism.

germination The beginning of plant growth that occurs when the seed coat cracks.

gibberellin One of a family of plant hormones that trigger the germination of seeds and interact with auxins in regulating growth and fruit development.

glycolysis The first stage of cellular respiration in all organisms that occurs in the cytoplasmic fluid, which results in a molecule of glucose splitting into two molecules of pyruvic acid.

Golgi apparatus Also known as Golgi complex. An organelle in eukaryotic cells consisting of stacks of membranous sacs that modify, store, and ship products of the endoplasmic reticulum.

grana A stack of hollow discs of thylakoid membranes in a chloroplast, the sites where light energy is trapped by chlorophyll and converted to chemical energy.

ground tissue system Parenchyma cells that make up the bulk of a young plant, and form continuous tissue throughout the body and fill the space between the epidermis and the vascular tissue system.

guard cell A specialized epidermal cell usually located on the underside of plant leaves that regulates the size of a stomata, allowing gas exchange between the surrounding air and the photosynthetic cells in the leaf.

half-life The period of time that it takes for one half of a given amount of a radioiostope to decompose into a more stable form.

haploid Having only one set of chromosomes, characteristic of gametes.

hemoglobin An iron-containing protein in red blood cells that transports oxygen from the lungs to body tissues.

heterotroph An organism that must get energy from outside food sources, not capable of making their own food.

heterozygous advantage Allows greater reproductive success of heterozygous individuals in a population as compared to homozygotes and tends to preserve variation in gene pools.

histones Various simple proteins stacked inside DNA molecules in the nucleus.

homologous structures Structures that share a common ancestry.

hormone A chemical messenger that travels in the blood from its production site, an endocrine gland, to target cells, which respond to the regulatory signal.

hydrogen bond A type of weak chemical bond between the partially positive charged hydrogen atom in a molecule to the partially negative charged atom in a different molecule.

hydrophilic Means "water loving"; refers to the molecular attraction of water.

hydrophobic Refers to the molecular repulsion of water; pertains to nonpolar molecules (or parts of molecules) that do not dissolve in water.

hypha One of many filaments making up the body of a fungus.

hypotonic solution In comparing two solutions, the one with lower concentration of solutes, and greater concentration of water.

incomplete dominance A type of inheritance where the expression of the genotype is a blending in the phenotype of a heterozygote: (Gg) is an intermediate expression between the homozygote's (GG and gg) phenotypes.

independent assortment A random distribution of homologous chromosomes during meiosis.

indoleacetic acid (IAA) A naturally occurring plant hormone.

inducer A substance that is capable of activating a structural gene by combining with and inactivating a repressor gene.

inhibitory proteins Specialized proteins that block a reaction or process.

internode The portion of a plant stem between two nodes.

interphase The longest period between two mitotic or meiotic divisions of a eukaryotic cell during which the cell carries out routine functioning, copies its DNA, and prepares to divide.

intertidal zone A productive, shallow zone where the waters of an estuary or ocean meet land.

intron A noncoding DNA sequence within a gene that is initially transcribed into messenger RNA.

ion An electrically charged atom or molecule, either positive or negative charge.

ion channels Specialized transport proteins located in cell membranes through which ions can pass without disturbance.

ionic bond A bond formed by electrical attraction between two oppositely charged ions.

isotonic solution A solution of equal solute concentration so that no net movement of water occurs.

karyotype A complete pictorial arrangement of the chromosomes found in an individual arranged in order of size and shape for identification.

Kreb's cycle A cellular process that includes a series of biochemical reactions that release carbon dioxide and create ATP.

light reaction The steps in which solar energy is absorbed and converted to chemical energy in the form of ATP and NADPH in the process of photosynthesis.

lignin A chemical compound that combines with cellulose to create the secondary cell wall in plants; lignin makes the cell wall more rigid.

linked genes Genes that are located close enough together on a chromosome so they are often inherited together.

lipid One of a family of nonpolar organic molecules that are not soluble in water; includes fats, phospholipids, and steroids.

littoral zone Highly productive, photosynthetic, shallow zone near the shore in a freshwater habitat.

lysogenic cycle A cycle in which a virus replicates in a host cell without destroying the host cell.

lysosome An organelle in a eukaryotic cell that contains digestive enzymes.

lytic cycle One type of viral infection that results in replication of virus and host cell destruction.

mating season The particular time of the year when a species prefers to reproduce; usually associated with an abundance of food for the offspring.

meiosis The process during cell division in which the nucleus of a cell completes two successive divisions that produce four gametes, each with a chromosome number that has been reduced by half.

meristematic tissue The only mitotic segment of a plant.

mesophyll The ground tissue of a leaf where most photosynthesis takes place.

microspheres Tiny spheres formed by short chains of amino acids.

mitochondrial matrix The fluid contained within the inner membrane of a mitochondrion where certain metabolic reactions occur.

mitosis The process also called cell division in which the nucleus of a cell divides into two nuclei, each with the same number and kind of chromosomes, which results in production of new cells for growth and repair.

monosaccharide A simple sugar that contains one sugar unit and is the base unit of carbohydrates.

multiple alleles The existence of more than two alleles for a genetic trait (blood types, for instance).

mutation A change in the DNA of a gene or chromosomes; the ultimate source of genetic variations.

mycelium The densely branched network of hyphae in a fungus.

mycorrhiza A mutualistic association of plant roots and fungal mycelium that creates greater absorption capacity for both organisms.

NADH An electron carrier that stores energy used to make ATP.

natural selection The process by which populations change in response to their environment as individuals better adapted to the environment generate more offspring than those individuals not suited to the environment.

nitrogen An element found in amino acids.

node The point of attachment of a leaf to a stem.

nondisjunction The failure of homologous chromosomes or sister chromatids to separate in meiosis or mitosis so that one daughter cell has both and the other neither of the chromosomes.

nuclear envelope Also called the nuclear membrane and composed of a double membrane that encloses the nucleus and separates it from the rest of the eukaryotic cell.

nucleic acid An organic molecule made of nucleotides that compose both DNA and RNA and store hereditary information for the cell.

nucleus An organelle that houses the DNA of eukaryotic cells and controls the function of the rest of the cellular organelles.

oligotrophic lakes Lakes that do not have a sediment or nutrient load.

oogenesis The process by which gametes (eggs) are produced in female animals.

osmosis The movement of water through a selectively permeable membrane from an area of high concentration to an area of lower concentration.

pangenes Incorrectly thought to be inheritable factors.

pangenesis A disproven hypothetical mechanism of heredity in which the body cells create pangenes that collect in the reproductive organs so that the egg or sperm contains particles from all body parts of the parent.

parenchyma cell A plant cell type that is a relatively unspecialized cell with a thin primary wall and no secondary wall; it functions in photosynthesis, food storage, and aerobic respiration, and may differentiate into other cell types.

passive transport The movement of a substance without the use of energy by the cell, usually along a concentration gradient.

pelagic zone The relatively unproductive area of an ocean occupied by deep open water.

phagocytosis A type of endocytosis whereby a cell engulfs macromolecules, other cells, or particles into its cytoplasm.

phenetics Relating to taxonomic analysis that emphasizes the overall similarities of characteristics among biological taxa without regard to phylogenetic relationships.

phenotype Observable characteristics of an organism.

pheromone Specific chemical signal produced by an organism that affects the behavior and/or development of other individuals of the same species.

phospholipid A lipid molecule in all membranes of a cell that contain phosphate units.

photic zone The productive region of an aquatic ecosystem into which light penetrates and where photosynthesis occurs.

photosynthesis The process by which green plants convert light energy into the chemical energy of organic compounds.

photosystems Two light-harvesting units found in every chloroplast's thylakoid membrane.

phylogenetic tree A branching diagram that represents evolutionary relationships among organisms.

pioneer effect The colonizing of an area by the first inhabitants.

plasmodesmata An open channel in a plant cell wall through which strands of cytoplasm connect from adjacent walls, thereby connecting the cells.

pleated sheets The folded arrangement of a polypeptide in a protein's secondary structure.

pleiotropy The control of more than one phenotypic characteristic by a single gene.

polar bodies The components of an ovule that do not become an egg, but develop into food for a fertilized egg.

polar covalent bond An attraction between atoms that share electrons unequally because the atoms differ in electronegativity; the shared electrons are pulled closer to the more electronegative atom, giving it a partial negative charge and the other atom a partial positive charge, creating an overall compound of a polar covalent nature.

polarity A measure of the charged area created when excess positive or negative charges predominate in an area.

pollination The delivery of pollen from the male parts of a plant to the stigma of a carpel on the female seed plant.

polygenic Traits characteristic of an organism that is influenced by several genes.

polymorphism The existence of multiple body forms, usually pertaining to a population in which two or more morphs are present in readily noticeable frequencies.

polyploid An unusual condition in which two or more complete sets of chromosomes exist within a cell.

polysaccharide A carbohydrate polymer consisting of multiple, even hundreds or thousands of monosaccharides (sugars) linked by covalent bonds.

primary growth The growth in the length of a plant root or shoot produced by an apical meristem, not a growth in girth.

primary structure The first level of protein structure that identifies the specific sequence of amino acids making up a polypeptide chain.

prokaryote A single-celled organism without a nuclear membrane such as a bacterial cell.

prophage Phage DNA that is inserted into the DNA of a prokaryotic chromosome, for genetic-engineering purposes.

proteins Amino acids joined together in specific sequences.

protist A member of the kingdom *Protista.*

pseudopodium A temporary extension of cytoplasm of the amoeba that can project in any direction and enables it to move.

punctuated equilibrium A model of evolution in which short periods of rapid change in a species are separated by long periods of little or no change.

quaternary structure The fourth level of protein structure in which the shape is determined by the union of two or more polypeptide units; functional protein configuration.

radiometric dating Dating of objects through the measurement and comparison of the half-life of certain radioisotopes and the products of their radioactive decay.

receptor A specific protein molecule located on or in a cell that has a specific shape that fits a specific messenger molecule, such as a hormone.

recessive A genetic trait that is not expressed unless the dominant allele of the trait is absent.

recombinant DNA DNA made from two or more different organisms.

regulatory gene A gene that regulates the expression of one or more structural genes by controlling the production of a protein that regulates their rate of transcription.

relative dating A technique used to determine the approximate age of fossils by comparing them with other fossils in different layers of rock with known strata linkages.

repressor A protein that binds to the operator in an operon to switch off transcription.

reproductive barriers Any biotic or abiotic factor that prevents successful reproduction.

reproductive isolation Prevention of mating between formerly interbreeding groups, or the inability of these groups to produce fertile offspring.

reverse transcriptase An enzyme that catalyzes the reverse process of transcription.

rhizobium A type of nitrogen-fixing soil bacteria.

rhizoid A hairlike projection that anchors a nonvascular plant.

ribosome A cellular organelle on which proteins are made.

RNA splicing The process by which intron segments of RNA are deleted.

RNA World Describes the hypothetical time of the earliest life-forms when genes were simply strands of RNA.

root burn A type of damage caused in root cells when the concentration of water in the soil draws enough water out of the root to injure the cells.

root pressure The upward push of xylem sap in a vascular plant, caused by the transpiration pull from the leaves and the active pumping of minerals into the xylem by root cells.

rough endoplasmic reticulum Endoplasmic reticulum cells that contain ribosomes.

sclerenchyma cells One of three types of plant cells with rigid secondary walls hardened with lignin.

secondary growth An increase in a plant's girth, involving cell division in the vascular cambium and the cork cambium; it is not a growth in length or height.

secondary structure The second level of protein configuration.

secretory protein A protein that is secreted by a cell, such as an antibody.

selectively permeable A membrane that allows some substances to pass through but not others.

sex-linked traits A trait that is determined by a gene found on the X chromosome, which differentiates between males and females.

sexual reproduction Reproduction in which gametes from opposite sexes or mating types unite to form a zygote.

sexual selection Traits that increase the ability of individuals to attract or acquire greater reproductive opportunities.

smooth endoplasmic reticulum Endoplasmic reticulum with no ribosomes attached.

spermatids One of the haploid cells that are formed by division of the secondary spermatocytes and that differentiate into sperm.

spermatogenesis The process by which gametes, sperm, are produced in male animals.

sporophyte The diploid phase that produces spores in the life cycle known as alternation of generations.

stabilizing selection A type of natural selection in which the intermediate form of the trait is favored and becomes more common, and extreme phenotypes are selected against.

stoma An opening usually in the underside of a leaf or a stem of a plant that enables gas exchange to occur and is controlled by guard cells that surround them.

substrate The substance on which an enzyme acts during a chemical reaction.

surface tension The attractive force exerted upon the surface molecules of a liquid by the molecules beneath that tends to draw the surface molecules into the shape having the least surface area; the measure of how much force is required to break the surface of a liquid.

sympatric Occupying the same territory without loss of identity from interbreeding.

systematic circulation The division of the circulatory system in which blood flows between the heart and all the tissues of the body other than the lungs.

target cell A specific cell that a hormone or enzyme binds to and acts on to produce a specific effect.

taxonomy The science of naming and classifying organisms.

test cross A means to determine the genotype of an individual whose phenotype is dominant but whose genotype is unknown by mating with a homozygous recessive individual and observing the phenotype of the offspring.

tetrad A group of four synapsed chromatids that become visibly evident in meiotic prophase.

thylakoids The internal membrane-bound sac in a chloroplast where photosynthesis occurs.

tissue A cooperative unit of many similar cells that perform a specific function within a multicellular organism.

transcription The stage of protein synthesis in which the information in DNA for making a protein is transferred to an RNA molecule.

transgenic An organism that contains foreign genes.

transgenic animal An animal that has had foreign DNA introduced into its cells.

translocation The stage of gene construction in which the information in mRNA is used to make a protein.

transpiration The evaporative loss of water from plant leaves, which increases root pressure.

tropical rain forest A biome in humid tropical areas where annual rainfall exceeds 98.5 inches or 250 cm and is the most complex and species-rich ecosystem.

true-breeding Displaying only one form of a particular trait in offspring for successive generations.

uniformitarianism A geological doctrine stating that existing processes act in the same manner as earlier geologic features.

universal solvent Water.

vaccine Substance prepared from killed or weakened pathogens and introduced into a body to produce a specific immunity to that pathogen.

vascular tissue A network of cells forming xylem tubes that extend throughout a plant, which channels water and minerals upward from the roots, and phloem, which channels sugars produced in the leaves throughout the plant.

vesicles Membrane-enclosed sacs in a cell's interior.

vessel element Water-conducting and supportive cell in plants combined with tracheids make up the water-conducting, supportive tubes in xylem.

zoomastigote A unicellular protist that is a member of the phylum *Zoomastigina.*

zygote A fertilized egg cell.

Index

A

B

D

E

F

G

H

I-J

K-L

M

N

O

P–Q

R

S

T

U-V

W

X-Y-Z